WHAT
FUTURE

AN UNNAMED PRESS BOOK

Copyright © 2018 Unnamed Press

All rights reserved, including the right to reproduce this book or portions thereof in any form whatsoever. Permissions inquiries may be directed to info@unnamedpress.com. Published in North America by the Unnamed Press.

www.unnamedpress.com

What Future, Unnamed Press, and the colophon, are registered trademarks of Unnamed Media LLC.

ISBN: 978-1-944700-66-9
eISBN: 978-1-944700-67-6

Catalog-In-Publication data available upon request.

This book is a work of nonfiction.

Designed & typeset by Jaya Nicely

Distributed by Publishers Group West
Manufactured in the United States of America
First Edition

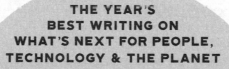

THE YEAR'S
BEST WRITING ON
WHAT'S NEXT FOR PEOPLE,
TECHNOLOGY & THE PLANET

WHAT
FUTURE

EDITED BY MEEHAN CRIST
& ROSE EVELETH

2018

TABLE OF CONTENTS

Introduction

By Meehan Crist
and Rose Eveleth

F uturism, with a capital F, is a technical practice. You can get fancy-sounding degrees in this practice, with names like "Master of Science in Foresight." You can then use those degrees to advise behemoth corporations like Coca-Cola or Bank of America on how they can make even more money in the future. But the practice of futurism, with a small f, is what most of us do when we daydream about what tomorrow might be. The act of imagining a different tomorrow might not seem like a radical practice, but it is. Or, it can be. Future thinking can be the first step in moving those dreams toward something like reality.

But in today's climate — both literally and figuratively — it can feel impossible to keep up with the current moment, let alone think about the future. Student debt weighs heavy on a generation poised to enter an increasingly precarious workforce. The safety nets of basic social services are being ripped out from beneath us. Becoming sick or injured can lead, with more alarming alacrity than ever, to financial ruin. Meanwhile, the planet burns. The waters are rising and crops are failing. The greatest human migration since the end of the first Ice Age is already underway, and borders around the world are tightening rather than opening up. Widespread species extinction is decimating biodiversity, and while the extinction of the human species seems unlikely, it is certainly not impossible. America is now engaged in perpetual global war, and nations around the world continue to suck fossil fuels from the ground even though we know plenty well where that will lead. The Pentagon has identified climate change as an immediate threat to global security, and yet we are well on our way to blowing past the goal of the Paris Accord to limit warming to 2 degrees Celsius, a ceiling rather flimsily calculated using "negative emissions" technologies that don't exist. Suffice it to say that for many Americans, 2017 was a dark year marked by extreme uncertainty about the future.

At the same time, there has been a perverse and pervasive cultural pressure to be optimistic — a nagging sense that optimism is some kind of moral good that chafes against the unending barrage of bad news. Former President Barack Obama likes to say, quoting Dr. Martin Luther King Jr., "The arc of the moral universe is long, but it bends toward justice," but there is little in recent or in deep human history to support a consoling Whig view of a progressive and inexorable movement toward a more just, humane, and sustainable world. History moves in far more complex motions than a single, inevitable arc. Similarly, there is no reason to believe that the unprecedented pace of technological development currently underway will lead us to a better world. Technological innovations are both created and deployed by within existing systems, where they get tangled up in mutually reinforcing webs, nearly always with unintended and unforeseen effects. How you feel about new technology may have more to do with your relation to those systems than whether you think you look silly in Oculus headgear.

In other words, how the future looks to you from the vantage point of the present depends largely on how you feel about the world we live in today. Perhaps you are a consumer whose clicking and liking has encouraged companies to bring you more of the products you like. Perhaps, because of this, you feel that having your online habits tracked and your data saved will lead to more of a good thing. Or perhaps you are someone for whom that data gathering has not gone quite so well, someone who has been flagged as "high-risk" for fraud or recidivism or crime. Perhaps you have been "singled out for punitive public policy and more intense surveillance" as Virginia Eubanks highlights in *Automating Inequality: How High-Tech Tools Profile, Police, and Punish the Poor.* If you experience what Eubanks calls "a kind of collective red-flagging, a feedback loop of injustice," then maybe a high-tech future powered by advanced data collection feels more dystopian than utopian.

Perhaps, if you are a person who feels they have "nothing to hide," the idea of the government using facial recognition software to track potential security threats is reassuring. Or, perhaps, if you are one of the American citizens recently tagged by the FBI as a "Black Identity

Extremist" for protesting police brutality, maybe facial recognition feels far less benign. If you are someone whose ancestry has never been used against them, the idea of 23andMe donating genetic kits to reunite the families whose children were stolen from their parents at the US border sounds like a generous offer. If, on the other hand, you come from a community whose bodies have been used by science to make profitable drugs without your knowledge or permission, the idea of handing your blood to a tech company might not sound like such a great idea.

Future thinking can, then, present a bit of a paradox. Many of the people who have the luxury of sitting around and opining about possible tomorrows are those who already have the comforts of time, money, and security. Like, say, Silicon Valley venture capitalists. It is precisely those who do not currently enjoy such luxuries who rely most on possibilities for the future, who make it through today by imagining that tomorrow could be different, could be better. In 2014, the NAACP estimated that 2.3 million incarcerated people were black Americans, making up a staggering 34 percent of the correctional population. Trans folks are murdered at a disproportionately high rate. Tomorrow, for many of us, is not guaranteed. And yet, or perhaps because of this, we still dream.

The future will be what we make it, for better or for worse. History will continue to unfold, shaped by the reverberative hum of a million choices large and small. From nuclear arms deals to what you eat for breakfast, the butterfly wings will flap. And as technological innovation barrels forward, and we interact with those innovations to help give shape to history, our futuristic daydreams will continue to collide with all sorts of ethical, political, and cultural questions. What does it mean to use CRISPR to alter the germ line of the human species? How should we use the natural resources currently floating around in space? As rising seas erode our coastlines, what sort of cities should we build, and why? To take steps toward a future that better serves us all, we will need to imagine and interrogate the possibilities with as much rigor and vision and heart as we can muster.

The scholar and activist adrienne marie brown often speaks of being locked in an "imagination battle." Like so many in a rich intellectual tradition that spans centuries and includes thinkers and writers from Karl Marx to Angela Davis to Kim Stanley Robinson, brown sees the creative work of mapping alternative futures in fiction and in fact as a potentially radical activity. "Sci-fi is simply a way to practice the future together," brown said in a speech in 2015, "I invite you to join me in writing ourselves into the future, naming the principles of total transformation, building an economy in which black lives matter because every single life, and all that supports life, matters—let us practice in every possible way the world we want to see." This is the work of futurism with a small f, and while visions of what "we" want will not always align, exploring those visions can help illuminate shared interests and suggest shared loci for something like hope.

In gathering the elements for this *What Future* collection, we were drawn to pieces that explore these ideas of practicing the future, naming principles of transformation, and writing folks into history who have previously been strategically written out. These are pieces that tell stories about future-facing experiments already underway, that try to name the phenomena that shape who we are and what we do, and that amplify marginalized voices too often silenced in the single-minded roar of the mainstream.

The articles and essays collected here offer some of the biggest imagining, the best analysis, and the most compelling writing we've read about the future. They have surprised, horrified, and energized us, and they have helped expand our imagination for what is possible.

To practice the future can mean rigorous imagining, but some people are trying to bring visionary plans into the here and now. Ross Anderson's piece, "Welcome to Pleistocene Park," introduces readers to a man practicing for the future by trying to recreate the past, attempting to meet the challenge of climate change by resurrecting the biome of the last Ice Age. Nikita Zimov lives in the remote forests of Eastern Siberia, where he is enacting sweeping plans that can only unfold in geological time, and thus require the commitment of future generations, including, perhaps most crucially, that of his own children.

Whether his vision is genius or folly, or something in between, remains to be seen.

Gavin Munro, the subject of Sarah Laskow's piece, "A Forest of Furniture Is Growing in England," is much like Zimov in that he's is trying to physically build an alternative future from the ground up. But rather than an entire ecosystem, Munro is focused on what might seem more banal, or even whimsical: chairs. He has, improbably, figured out a way to grow chairs by coaxing young trees to mature into the shape of furniture. In doing so, he's hoping to shape a future in which people rethink the nature of objects themselves: where they come from and what they should, or could, be. Like Zimov, Munro is working in tune with geological timescales over which he ultimately has no control. "To reshape the world means believing in the promise of sweeping, far-sighted plans, and understanding that two millennia might only be half a lifetime," Laskow writes. These grand plans have impacts on shorter timelines too: even if Munro's geological projects never take root, creating a farm and building a community are actions that will affect the future in ways that no one can predict. From the seed of an idea, the future grows.

While some futurists may want to remake the planet, others help us name principles of transformation. Some of the best writing about the future illuminates a process that is shaping our world, but isn't up on a TED stage declaring itself to be doing so. Ross Exo Adams's essay, "Becoming-Infrastructural," explores how the built environment controls our bodies and circumscribes political possibilities, revealing the nature of buildings and cities in a way that most of us probably haven't considered, despite interacting with this infrastructure every day. Adams takes a close look at "resilience architecture," a concept now in vogue among architects and city planners looking for ways to create human habitats along threatened coastlines, and reminds us that as we choose how to respond to climate change, we may be shaping the future in unexpected ways.

Hidden forces that need naming also abound in the digital realm, and Ed Finn's "How Predictions Can Change the Future" names another process hidden in plain sight. We dare you to look at Google the same way after reading his analysis of how each search impacts what your feed and autocompleted search terms look like the next time you start typing into that seemingly innocuous little rectangle. Finn reveals how technology we interact with every day subtly shifts the way we act in the future, and how the "futures you can think about are shaped by somebody else's code."

Future thinking can destabilize even what is most familiar, but it can also provide a false sense of security and control. As Olivia Rosane highlights in "Breaking the Waves," the way we draw climate change maps can make the future feel knowable, mappable, and exact. "Don't worry, the graphics say, even if the water rises, it will rise in a neat and orderly progression." On the ground, sea level rise promises to be anything but, and this illusion of order may affect how we act in the face of rising seas.

Our tendency to map the unknown into plausible and reassuring order pops up in medicine, as well. In "The Eaten World," gastroenterologist Nitin K. Ahuja considers recent enthusiasm for the microbiome and examines how the habit of mapping our bodies to the world as we know it changes the way we think about human health and medicine. It is fashionable to think of the microbiome as a wounded ecosystem, he argues, partly because we live in a time of threatened environments, but he questions whether this parallel is medically meaningful, and how this reading might fail us when we know so little about how the microbiome actually works.

Sometimes, however, we know almost exactly what the future will bring and how we will handle it. There are protocols. Plans in place. Sam Knight's piece, "London Bridge is Down": The Secret Plan for the Days After the Queen's Death," details the byzantine theatrics that will kick into high gear not if, but when the Queen of England dies. He wonders just what all this theatre is for, exactly. "Knowing everything that we know in 2017, how can it possibly hold that a single person might contain the soul of a nation?" Perhaps it is time to leave old structures behind and for new forms of civic theatre, and new politics, to emerge?

The old structures of historical bureaucracy and colonialism are reaching forward into the future whether we like it or not. They are even reaching into outer space. Atossa Araxia Abrahamian's piece,

"How a Tax Haven is Leading the Race to Privatize Space" shows how Silicon Valley venture capitalists are teaming up with the tiny country (and financial powerhouse) of Luxembourg to privatize space, or as they call it, "new space." Together, they're building futures markets for space mining, and making big bets on what materials extracted from asteroids might someday be worth. In the process, they're conveniently bypassing any international discussion about those resources, which the Outer Space Treaty of 1967 internationally ratified as shared property, a universe to be shared by all humanity.

Sometimes it can seem easier to imagine mining asteroids than to imagine systemic change back down on Earth. But that kind of imagination is crucial to our future. In "What Will Kill Neoliberalism?" five authors tackle that very question in brief but intriguing flashes, offering five different (though not mutually exclusive) answers. Bryce Covert posits that neoliberalism will crumble under the weight of the crisis of care—the increasing inability for wage-earners to both make ends meet and care for their children, the sick, and the elderly —which is already fostering solidarity among millions of Americans and pushing people to resist prevailing economic conditions. Paul Mason sees it ending with the end of globalism, the doomed rise of economic nationalism, and the eventual transition to a low-work, high-abundance society. Peter Barnes argues that universal basic income will shift us to a new economic and social model. William Darity predicts that managers—a class composed of intellectuals, artists, artisans, and state bureaucrats, who are positioned between the working class and owners of capital—will ultimately have to pick a side, and just might be the straw that breaks neoliberalism's back. Joelle Gamble points to populist responses to neoliberalism on both the right and left, sketching out how populism from the right could usher in a fully privatized and authoritarian state, while populism from the left could lead to a mass redistribution of wealth and power. The ascendancy of neoliberalism and the pull of the markets can feel like the workings of natural laws, woven into the fabric of spacetime, and neoliberalism has been the tent under which Americans have been living for as long as most of us can remember. But history tells us that any society marked by an increasingly unequal distribution of wealth eventually ends in some form of economic and social collapse. These five authors remind us that the future of our economic and political structures are not preordained, nor are they driven by inexorable forces. They're created by humans and they can be changed or destroyed by us, too. The question of systemic overhaul has never really been "if", but rather "when" and "how."

One concept proposed in "What Will Kill Neoliberalism?", universal basic income, or UBI, bubbled to the surface of our collective imagination (again) in response to deepening poverty and the ever-increasing automation of industries around the world. Brishen Rogers's piece, "Basic Income in a Just Society," offers a thoughtful exploration of not just how a guaranteed income might work but also, crucially, a lucid exploration of "the policy's justifications, merits, and limits," including libertarian and leftist visions of UBI and whether the problems many think it will solve even exist. Rogers, a lawyer and former community organizer, writes, "The reports of the death of work have been greatly exaggerated... But we still need a vision of good work and its place in our society, one that recognizes how our economy—and our working class—have changed dramatically in recent decades." The capitalist structures of workers and bosses are being reinforced and strengthened by automation, which creates additional layers of data that put new pressures onto workers. This is just one of the problems basic income has to be able to solve, Rogers argues, if such a policy is truly going to make the future better.

While a national UBI or actual asteroid mining may still be a long way off, in some places it can seem as if the future has already arrived. Consider Estonia. This Baltic nation has a virtual government, electronic voting, blockchained security, borderless citizenship, digital health records, and lawn-mowing robots. "It was as if, in the nineties, Estonia and the U.S. had approached a fork in the road to a digital future, and the U.S. had taken one path—personalization, anonymity, information privatization, and competitive efficiency—while Estonia had taken the other," writes Nathan Heller in his piece "Estonia, the Digital Republic." The government of Estonia argues that this system saves money and time, but it's also a savvy geopolitical strategy: when Russia comes for them, they will have everything backed up.

They will be ready, not just to save their documents, but to run their country remotely, from the cloud.

Meanwhile, 2,500 miles away, in Kashgar, China, another future has already emerged: a sprawling system of surveillance technologies watches citizens' every move. In "This Is What a 21st-Century Police State Really Looks Like," Megha Rajagopalan reveals the claustrophobic ways that new technologies are being used to surveil and police the region's Uighur minority on a daily basis. The scariest part of Rajagopalan's piece, however, isn't the technology. It's the systems into which that technology lands, amplifying state control of an already oppressed minority. While it may be easier for Americans to see dystopia when we look to countries like China, a future of constant surveillance is already a reality in the United States. As we were writing this introduction, we both received emails from the ACLU about how Amazon is providing its facial recognition technologies to governments and law enforcement, quite likely for use in your city. If you happen to be poor in America, the State already watches what you buy with your EBT card, polices your relationships with your children, and second-guesses your need for medical care.

Perhaps the most inevitable thing about the future is that we're all going to age and die. So often future thinking sets aside the fact that old people will continue to exist, or assumes that soon we're going to live forever, but as the Boomer generation grows older, we are facing a future with the largest aging population America has ever seen, what has somewhat contentiously been dubbed the "silver tsunami." While this language suggests a massive surge of incapacitated elderly, and thus fails to reflect the myriad ways people live into old age, we can safely say that older people need different kinds of care and we're currently facing a future where that care will be harder to come by. In "Someone to Watch Over Me: What Happens When We Let Tech Care for our Aging Parents" Lauren Smiley grapples with the question of how much we want to hand our seniors over to technology. Can a cat avatar powered by AI and humans-for-hire, who are watching an iPad camera feed from the other side of the planet, really provide enough support for an aging loved one? And even if it can, does that feel right?

Or how about the power to control human memory? If advancements in neuroscience were to make it possible, would that be something we want? In "The Future of Remembering," Rachel Riederer talks with researchers doing science that might lead to conquering dementia and PTSD, and could even enhance healthy memory function. New methods of memory control are years from going to market, but interest can be counted to the tune of millions of dollars. "As human memory changes from an intractable mystery to something that can be engineered," Riederer asks, "who will get to decide how it works?"

Many of the pieces in this collection grapple with hope and despair, or their conceptual cousins, optimism and pessimism. Last year, when David Wallace-Wells published a decidedly pessimistic piece on the future of our planet in the grip of climate change in *New York Magazine*, taking worst-case scenarios seriously and using climate science to sketch them out, the response was explosive. People's reactions were so strong in part because he got some of the science wrong, but also because his vision was so very bleak. What ensued was a fascinating (and highly Google-able) debate about "the right way" to talk about climate change. One thoughtful response reprinted here, "Why Hope Is Dangerous When It Comes to Climate Change," by Tommy Lynch, offers a compelling argument for how we might go about balancing hope, optimism and despair in times that are inarguably dark.

Perhaps one of the darkest features of the United States today is pervasive gun violence. In "Magic Bullets," Patrick Blanchfield considers people's reactions to 3-D printed guns and interrogates "our collective fascination with gun futurism—our reactions, variously hopeful or hysterical, utopian or bleak." What he finds is that sometimes looking to the past is the only way to see clearly into the future. Addressing the particularly American crisis of gun violence, he argues, requires looking long and hard at the entwined relations between capitalism and American militarism. "The real issue isn't our fantasies about the future," he writes, "but how focusing on them helps us ignore the legacy of our past and the realities of our present."

Lidia Yuknavitch's piece, "Civic Memory, Feminist Future," also looks back to look forward, mining her own history as a woman and an artist in an attempt to illuminate a path forward for feminism.

"Seventeen is the cusp of everything," she writes. "A girl's mind morphs toward woman faster than her brain can track, and so her body lurches and grinds forward more like an animal's. All around her images from her culture of what to be, what to look like, how to be wanted and thus counted." Her essay serves as a rigorous, intimate reminder of how art is inextricable from politics, and how under patriarchy the personal will always be political. "What would a realized woman look like in America?" she asks. "What is history?... Do we move within it, or against it, or do we merely view it, consume it and move on?... What part is ours to move?"

In a similar vein, but a different register altogether, Walidah Imarisha's "Science Fiction, Ancient Futures, and the Liberated Archive" engages with the hard work of writing ourselves into both the past and the future. A lightly edited transcript of a keynote address she gave at the Society of American Archivists 2017 Annual Conference, her piece rolls out a roadmap for a world in which marginalized people are not strategically forgotten in our shared narratives. Thinking about the past as savagery, she argues, is a colonial way of seeing the future. "I think it's really important to recognize that linear time is a method of social control," Imarisha writes. "We're told this is all we have. The past is gone, there's nothing it can do for you. The future is unknowable, you can't do anything about that." But, she points out, "many societies, historically...you know, brown cultures have recognized that time is not linear, it's circular, it's spherical, it exists in multiple places at once, that we live in the past, the present, and the future altogether." Drawing on examples from the almost-erased history of black people in Oregon, she invites her audience to imagine a world where traditionally marginalized perspectives help shape our collective future. Her own desire for such a world is palpable: "I want that future so badly, y'all."

Desire is fundamentally about the future—wanting what one does not have... yet. The word "desire" appears 21 times in Andrea Long Chu's essay, "On Liking Women," in which Chu interrogates her own experience of yearning in order to unravel the way trans women can feel tangled up by cultural narratives and theoretical arguments about who should want what and why. Any piece that centers trans experience at this moment in history is inherently future-facing, and Chu explicitly engages feminist theory that imagines future sexualities and genders written by feminist thinkers such as Valerie Solanas, author of the 1967 radical feminist text *SCUM Manifesto*. Chu writes: "all cultural things, *SCUM Manifesto* included, are answering machines for history's messages at best only secondarily. They are rather, first and foremost, occasions for people to feel something: to adjust the pitch of a desire or up a fantasy's thread count, to make overtures to a new way to feel or renew their vows with an old one. We read things, watch things, from political history to pop culture, as feminists and as people, because we want to belong to a community or public, or because we are stressed out at work, or because we are looking for a friend or a lover, or perhaps because we are struggling to figure out how to feel political in an age and culture defined by a general shipwrecking of the beautiful old stories of history." She reminds us that reaching for utopia requires taking imperfect steps towards what may be impossible desires.

In an age defined by "a general shipwrecking of the beautiful old stories of history," looking to the future means listening carefully to young people who will write the beautiful new stories (for shipwrecking, perhaps, at some later date). Bryan Washington's piece, "The Future in Motion: Why I Judge High School Debate Tournaments" is an intimate look inside the world of high school debates as experienced by young students of color. Washington watches kids stand on chairs or dramatically collapse to make their points; he watches them perform against white peers at far fancier schools with brilliance and bravery; he watches how they make the judges hold their breath, and wonders what will become of them. "On any given weekend in Alief, you'll end up judging a room full of Latino kids, all of them fluent in the "One China" policy. Or a roster that's half-Nigerian, riffing on public school funding. Or twenty Vietnamese kids poking at cyber security...It is science fiction, in a literal sense," Washington writes. "It's also fucking absurd. And it's as clear a portrait of our nation as any you're likely to find." In Washington's piece we find possible reason for hope in dark times: these kids will help shape a world that none of us can possibly imagine today.

New worlds and new stories require new methods, and in "The Comparative Method: A Novella," Gretchen Bakke offers readers just such a method in a piece that challenges description in the most delightful way. This essay blurs the boundaries of nonfiction and fiction, academia and art, by clearly commenting on our present reality through a powerfully imagined voice from the future. "There comes a time in a woman's life when she begins to think about who will care for the machines she has left behind," begins Bakke, slowly building a world similar to, but not quite the same as, our own and describing a boxlike machine called The Comparative Method. "I had thought that The Comparative Method might move us back, toward a world of inclusion, toward a more perfect meritocracy... The Comparative Method chose the other route. It brought down the rankers. It unmade them; it anticoagulated them... It could whistle the pigeons into docility; it could hack door locks and photocopiers (but nothing else)..." Rather than attempt to "explain" the method(s) of Bakke's piece, we'll preserve the pleasure of discovering them for yourself.

The future will always be more terrible and wonderful than we can possibly imagine. It is precisely this uncertainty that inspires both dread and the possibility of hope. As the incomparable Ursula K. LeGuin once wrote, "The only thing that makes life possible is permanent, intolerable uncertainty: not knowing what comes next." But when we allow ourselves to imagine beyond the present tense—the current political structures, technologies, and social fictions that shape the world(s) we inhabit — we are often reaching for a future we want, and just as often trying to articulate those futures we are willing to stand against.

The pieces collected here might chafe against the kind of techno-optimistic futurism that you're used to, perhaps that you even picked up this book hoping to read. But without positioning the future in relation to the grim realities of today, acknowledging problems as well as potential, no quantity or quality of imagination will make life in the future any more just, or humane, or bearable for us or those who will come after.

Which is not to say any of us will get what we think we want. As Andrea Long Chu writes, "You don't want something because wanting it will lead to getting it. You want it because you want it. This is the zero-order disappointment that structures all desire and makes it possible. After all, if you could only want things you were guaranteed to get, you would never be able to want anything at all." And so we dream. We try.

WHAT
FUTURE

How Predictions Can Change the Future

By Ed Finn

(Originally appeared in
Slate, September 2017)

C an computers predict the future? We desperately want them to, if you count the sheer tonnage of science-fiction tales we've consumed over the decades featuring all-knowing techno-oracles using their massive calculating power to work out every detail in the same way IBM's Deep Blue games out a chess match. The magnificent Minds modeling the behavior of entire civilizations while calculating hyperspace jumps in Iain M. Banks' *Culture* novels. C-3PO rattling off the odds of survival to Han Solo in *Star Wars*.

For now, however, silicon seers aren't prophesizing the distant future like A.I. gods. Instead, they're creeping into the near future, gradually extending the reach of what computer engineers variously call foresight, anticipation, and prediction. Autonomous cars slam on the brakes seconds before an accident occurs. Stock-trading algorithms foresee market fluctuations crucial milliseconds in advance. Proprietary tools predict the next hit pop songs and Hollywood movies. The ways in which computation is sidling up to the future reminds me of the old William S. Burroughs line: "When you cut into the present, the future leaks out."

Better near-future predictions are beginning to appear in all sorts of consumer products, too. We used to laugh at wacky Amazon recommendations and Microsoft's infamous Clippy popping up to ineffectually "help" you write a letter in Word, but the predictions we see these days more often feel eerily accurate.

Consider Google autocomplete, those helpful little strings of suggested text that pop up as you start typing in a search query. As a genre of prophecy, this might seem pretty lame. But consider how often the typical internet user relies on those little pop-ups every day, using them not just to save typing a few more letters but as a kind of microquery in its own right: a rapid spell-check, fact check, and zeitgeist check all rolled into one. Does the name you're searching for pop up with "girlfriend" or "married" appear after it? What does Google suggest after you type in "how do i"? (Google's suggestions for me are "get home; renew a passport; get a passport; love thee." Thanks, Google.) These predictions are based on the words thousands of other people typed into their search bars, but they are also customized for you based on your own browsing history, location, and whatever else Google might care to reference in its extensive file on you.

This may seem like an unremarkable convenience, but it is also a way to reinvent the relationship we all have with "now" and "soon." Years ago, Google realized that people get annoyed by delays in response time—even if it's less than a second. In fact, humans can be bothered by any lags that are perceptibly longer than the speed at which our own nervous system can respond to stimuli (about 250 milliseconds). So if Google wants to get you something *now*, it strives to do it in about the time it takes for your foot to report that you have stubbed your toe *just now*. In doing so, it pushes the envelope of instant gratification by attempting to guess what you want before you even articulate it. Autocomplete leaps ahead of now to offer you the near future on a silver platter.

The predictive quality of the algorithm gets more interesting when you pick an autocomplete suggestion that wasn't quite what you were searching for but was close enough that you went along with it because you are a lazy mammal. For in this matter of your query about cats or clown anxieties, Google has not just predicted the future but changed it. Now multiply that possibility by the 3.5 billion or more search queries the company processes each day.

Autocomplete provides just one small example of the many ways algorithms' predictions shape the future. Think about your relationship with Facebook: the primary source of news for many people. The social network has deployed extensive predictive resources to figure out how to populate your feed with content and connections that will keep you coming back to the site. Are you liberal or conservative? Rich or poor? What's your ethnicity, your geographical location, your favorite brand of clothing? Advertisers also want to know—and have been flocking to Facebook's increasingly formidable abilities to hone in on your demographics, influences, and preferences. As the ads become better targeted, they're more likely to influence the products you purchase, gifts you get, trips you take, neighborhood you move to, or when you make major decisions such as changing jobs or getting married. And the algorithms' decisions about who to share your next big life event with continue the feedback loop for others. That data may also be used to discriminate against you. It may have already done so, as it did with an old Facebook advertising system that allowed clients to exclude particular "ethnic affinities" from seeing housing, credit, and employment ads. (The company now claims it's enacting policies to prevent this kind of deliberate bias.)

Even if there's no large manila folder in Palo Alto labeled "black people on Facebook," the algorithms are written to differentiate perceived demographics in an instrumental way. The program will populate your feed with posts and ads it thinks you—or rather, its sometime uncanny but ultimately imprecise understanding of you— will enjoy, each with the potential to become a self-fulfilling prophecy of what you will like. The feed looks similar for those it thinks are similar to you but very different for someone it associates with other demographic and social categories. This is yet another kind of future-shaping: It manipulates the information we're aware of—not just the things we concentrate on, but the things on the horizon we are vaguely aware of. But those things on the margins often come back and become the future. Since we have only so much capacity to think about the decisions that aren't in the immediate now, catching a glimpse of something out of the corner of the eye—say a promo for a sneaker company your friends on Facebook like—can influence you later because the ad puts those kicks onto the fairly short list of things you might think about wanting later. It also makes the decision to buy sneakers from that company easier for you (a fact advertisers know full well). Our algorithms snuggle right in there between our complacency and our anxiety to fill the empty places. That ad for a realtor or an engagement ring might be close enough to the thing we thought we wanted that it's easier just to click on it. But by doing so, we become the flattened versions of ourselves that the algorithm predicted, and along the way, limit the possibilities of our experience. Lazy mammals.

"What if your brain, an incredibly adaptable tool, reshapes its thinking to better suit the algorithm, and you find yourself thinking in Facebook..."

This kind of prediction will only become more prevalent, and more seductive in its convenience, as algorithms improve and we feed them more data with our queries, our smart home devices, our social media updates, and our expanding archives of photos and videos. Facebook, for one, is beavering away at literally *reading your mind*. At its annual developer conference, it unveiled a new technology that can transcribe text directly from thought. It made for a very cool demo, but it also opened up a host of questions. What if, like autocomplete, the text is *almost* what you thought? What if your brain, an incredibly adaptable tool, reshapes its thinking to better suit the algorithm, and you find yourself thinking in Facebook, like you dreamed in French for a few weeks before the big Advanced Placement exam? Now the futures you can think about are shaped by somebody else's code.

This is an extreme version of how computers might predict the future, at least in its particulars. But in its generalities, it's happening all the time. We slice and dice the present in millions of different models, and making all sorts of assumptions about what we can and can't predict. The more we depend on computers to handle the near-future for us (where do I turn? what should I read? who should I meet?), the more limited our map of the present and potential future becomes. We're reversing William S. Burroughs: cutting up the future and turning it into discrete pieces of the present that have been denatured of ambition, of mystery, of doubt, and of deeper human purpose. We may get an answer to our question, but we don't know what it means.

For most of us, most of the time it's not the long-term, hazy-outline future that matters. It's the next five minutes, the next day, the next line of conversation. These are the predictions that algorithms want to make for us because they shape the real decisions that move our lives forward. But the future isn't just about the coming decisions in our field of view. It's the blank space on the map, the zone of possibility and hope. On good days, it's the telescope through which we see our best selves finally coming into being. But these algorithms can be so compelling in the ways that end up mapping out our near future that they obscure the slow, dramatic changes that might take decades to pull off. Algorithms are so effective at filling every available moment of free time with pings and updates that they foreclose vital opportunities for daydreams and self-reflection: the times when you suddenly realize you have to quit your job or write a book or change who you are. We've all had that experience of discovering a whole new version of ourselves totally unpredicted by any model. Those futures are not something we should give up to an algorithm.

Welcome to Pleistocene Park

By Ross Andersen

(First appeared in the *Atlantic*, April 2017)

N ikita Zimov's nickname for the vehicle seemed odd at first. It didn't look like a baby mammoth. It looked like a small tank, with armored wheels and a pit bull's center of gravity. Only after he smashed us into the first tree did the connection become clear.

We were driving through a remote forest in Eastern Siberia, just north of the Arctic Circle, when it happened. The summer thaw was in full swing. The undergrowth glowed green, and the air hung heavy with mosquitoes. We had just splashed through a series of deep ponds when, without a word of warning, Nikita veered off the trail and into the trees, ramming us into the trunk of a young 20-foot larch. The wheels spun for a moment, and then surged us forward. A dry crack rang out from under the fender as the larch snapped cleanly at its base and toppled over, falling in the quiet, dignified way that trees do.

I had never seen Nikita happier. Even seated behind the wheel, he loomed tall and broad-shouldered, his brown hair cut short like a soldier's. He fixed his large ice-blue eyes on the fallen tree and grinned. I remember thinking that in another age, Nikita might have led a hunter-gatherer band in some wildland of the far north. He squeezed the accelerator, slamming us into another larch, until it too snapped and toppled over, felled by our elephantine force. We rampaged 20 yards with this same violent rhythm—churning wheels, cracking timber, silent fall—before stopping to survey the flattened strip of larches in our wake.

"In general, I like trees," Nikita said. "But here, they are against our theory."

Behind us, through the fresh gap in the forest, our destination shone in the July sun. Beyond the broken trunks and a few dark tree-lined hills stood Pleistocene Park, a 50-square-mile nature reserve of grassy plains roamed by bison, musk oxen, wild horses, and maybe, in the not-too-distant future, lab-grown woolly mammoths. Though

"Were that frozen underground layer to warm too quickly, it would release some of the world's most dangerous climate-change accelerants into the atmosphere..."

its name winks at *Jurassic Park*, Nikita, the reserve's director, was keen to explain that it is not a tourist attraction, or even a species-resurrection project. It is, instead, a radical geoengineering scheme.

"It will be cute to have mammoths running around here," he told me. "But I'm not doing this for them, or for any other animals. I'm not one of these crazy scientists that just wants to make the world green. I am trying to solve the larger problem of climate change. I'm doing this for humans. I've got three daughters. I'm doing it for them."

Pleistocene Park is named for the geological epoch that ended only 12,000 years ago, having begun 2.6 million years earlier. Though colloquially known as the Ice Age, the Pleistocene could easily be called the Grass Age. Even during its deepest chills, when thick, blue-veined glaciers were bearing down on the Mediterranean, huge swaths of the planet were coated in grasslands. In Beringia, the Arctic belt that stretches across Siberia, all of Alaska, and much of Canada's Yukon, these vast plains of green and gold gave rise to a new biome, a cold-weather version of the African savanna called the Mammoth Steppe. But when the Ice Age ended, many of the grasslands vanished under mysterious circumstances, along with most of the giant species with whom we once shared this Earth.

Nikita is trying to resurface Beringia with grasslands. He wants to summon the Mammoth Steppe ecosystem, complete with its extinct creatures, back from the underworld of geological layers. The park was founded in 1996, and already it has broken out of its original fences, eating its way into the surrounding tundra scrublands and small forests. If Nikita has his way, Pleistocene Park will spread across Arctic Siberia and into North America, helping to slow the thawing of the Arctic permafrost. Were that frozen underground layer to warm too quickly, it would release some of the world's most dangerous climate-change accelerants into the atmosphere, visiting catastrophe on human beings and millions of other species.

In its scope and radicalism, the idea has few peers, save perhaps the scheme to cool the Earth by seeding the atmosphere with silvery mists of sun-reflecting aerosols. Only in Siberia's empty expanse could an experiment of this scale succeed, and only if human beings learn to cooperate across centuries. This intergenerational work has already begun. It was Nikita's father, Sergey, who first developed the idea for Pleistocene Park, before ceding control of it to Nikita.

The Zimovs have a complicated relationship. The father says he had to woo the son back to the Arctic. When Nikita was young, Sergey was, by his own admission, obsessed with work. "I don't think he even paid attention to me until I was 20," Nikita told me. Nikita went away for high school, to a prestigious science academy in Novosibirsk, Siberia's largest city. He found life there to his liking, and decided to stay for university. Sergey made the journey to Novosibirsk during Nikita's freshman year and asked him to come home. It would have been easy for Nikita to say no. He soon started dating the woman he would go on to marry. Saying yes to Sergey meant asking her to live, and raise children, in the ice fields at the top of the world. And then there was his pride. "It is difficult to dedicate your life to someone else's idea," he told me.

But Sergey was persuasive. Like many Russians, he has a poetic way of speaking. In the Arctic research community, he is famous for his ability to think across several scientific disciplines. He will spend years nurturing a big idea before previewing it for the field's luminaries. It will sound crazy at first, several of them told me. "But then you go away and you think," said Max Holmes, the deputy director of Woods Hole Research Center, in Massachusetts. "And the idea starts to makes sense, and then you can't come up with a good reason why it's wrong."

Of all the big ideas that have come spilling out of Sergey Zimov, none rouses his passions like Pleistocene Park. He once told me it would be "the largest project in human history."

As it happens, human history began in the Pleistocene. Many behaviors that distinguish us from other species emerged during that 2.6-million-year epoch, when glaciers pulsed down from the North Pole at regular intervals. In the flood myths of Noah and Gilgamesh, and in Plato's story of Atlantis, we get a clue as to what it was like when the last glaciation ended and the ice melted and the seas welled up, swallowing coasts and islands. But human culture has preserved no memory of an *oncoming* glaciation. We can only imagine what it was like to watch millennia of snow pile up into ice slabs that pushed ever southward. In the epic poems that compress generations of experience, a glaciation would have seemed like a tsunami of ice rolling down from the great white north.

One of these 10,000-year winters may have inspired our domestication of fire, that still unequaled technological leap that warmed us, warded away predators, and cooked the calorie-dense meals that nourished our growing brains. On our watch, fire evolved quickly, from a bonfire at the center of camp to industrial combustion that powers cities whose glow can be seen from space. But these fossil-fueled fires give off an exhaust, one that is pooling, invisibly, in the thin shell of air around our planet, warming its surface. And nowhere is warming faster, or with greater consequence, than the Arctic.

Every Arctic winter is an Ice Age in miniature. In late September, the sky darkens and the ice sheet atop the North Pole expands, spreading a surface freeze across the seas of the Arctic Ocean, like a cataract dilating over a blue iris. In October, the freeze hits Siberia's north coast and continues into the land, sandwiching the soil between surface snowpack and subterranean frost. When the spring sun comes, it melts the snow, but the frozen underground layer remains. Nearly a mile thick in some places, this Siberian permafrost extends through the northern tundra moonscape and well into the taiga forest that stretches, like an evergreen stripe, across Eurasia's midsection. Similar frozen layers lie beneath the surface in Alaska and the Yukon, and all are now beginning to thaw.

If this intercontinental ice block warms too quickly, its thawing will send as much greenhouse gas into the atmosphere each year as do all of America's SUVs, airliners, container ships, factories, and coal-burning plants combined. It could throw the planet's climate into a calamitous feedback loop, in which faster heating begets faster melting. The more apocalyptic climate-change scenarios will be in play. Coastal population centers could be swamped. Oceans could become more acidic. A mass extinction could rip its way up from the plankton base of the marine food chain. Megadroughts could expand deserts and send hundreds of millions of refugees across borders, triggering global war.

"Pleistocene Park is meant to slow the thawing of the permafrost," Nikita told me. The park sits in the transition zone between the Siberian tundra and the dense woods of the taiga. For decades, the Zimovs and their animals have stripped away the region's dark trees and shrubs to make way for the return of grasslands. Research suggests that these grasslands will reflect more sunlight than the forests and scrub they replace, causing the Arctic to absorb less heat. In winter, the short grass and animal-trampled snow will offer scant insulation, enabling the season's freeze to reach deeper into the Earth's crust, cooling the frozen soil beneath and locking one of the world's most dangerous carbon-dioxide lodes in a thermodynamic vault.

To test these landscape-scale cooling effects, Nikita will need to import the large herbivores of the Pleistocene. He's already begun bringing them in from far-off lands, two by two, as though filling an ark. But to grow his Ice Age lawn into a biome that stretches across continents, he needs millions more. He needs wild horses, musk oxen, reindeer, bison, and predators to corral the herbivores into herds. And, to keep the trees beaten back, he needs hundreds of thousands of resurrected woolly mammoths.

As a species, the woolly mammoth is fresh in its grave. People in Siberia still stumble on frozen mammoth remains with flesh and fur intact. Some scientists have held out hope that one of these carcasses may contain an undamaged cell suitable for cloning. But *Jurassic Park* notwithstanding, the DNA of a deceased animal decays quickly. Even if a deep freeze spares a cell the ravenous microbial swarms that follow in death's wake, a few thousand years of cosmic rays will reduce its genetic code to a jumble of unreadable fragments.

You could wander the entire Earth and not find a mammoth cell with a perfectly preserved nucleus. But you may not need one. A mammoth is merely a cold-adapted member of the elephant family. Asian elephants in zoos have been caught on camera making snowballs with their trunks. Modify the genomes of elephants like those, as nature modified their ancestors' across hundreds of thousands of years, and you can make your own mammoths.

The geneticist George Church and a team of scientists at his Harvard lab are trying to do exactly that. In early 2014, using CRISPR, the genome-editing technology, they began flying along the rails of the Asian elephant's double helix, switching in mammoth traits. They are trying to add cold-resistant hemoglobin and a full-body layer of insulating fat. They want to shrink the elephant's flapping, expressive ears so they don't freeze in the Arctic wind, and they want to coat the whole animal in luxurious fur. By October 2014, Church and his team had succeeded in editing 15 of the Asian elephant's genes. Late last year he told me he was tweaking 30 more, and he said he might need to change only 50 to do the whole job.

When I asked Beth Shapiro, the world's foremost expert in extinct species' DNA, about Church's work, she gushed. "George Church is awesome," she said. "He's on the right path, and no one has made more progress than him. But it's too early to say whether it will take only 50 genes, because it takes a lot of work to see what each of those changes is going to do to the whole animal."

Even if it takes hundreds of gene tweaks, Church won't have to make a perfect mammoth. If he can resculpt the Asian elephant so it can survive Januarys in Siberia, he can leave natural selection to do the polishing. For instance, mammoth hair was as long as 12 inches, but shorter fur will be fine for Church's purposes. Yakutian wild horses took less than 1,000 years to regrow long coats after they returned to the Arctic.

"The gene editing is the easy part," Church told me, before I left for Pleistocene Park. Assembling the edited cells into an embryo that survives to term is the real challenge, in part because surrogacy is out of the question. Asian elephants are an endangered species. Few scientists want to tinker with their reproductive processes, and no other animal's womb will do. Instead, the embryos will have to be nurtured in an engineered environment, most likely a tiny sac of uterine cells at first, and then a closet-size tank where the fetus can grow into a fully formed, 200-pound calf.

No one has yet brought a mammal to term in an artificial environment. The mammalian mother–child bond, with its precisely timed hormone releases, is beyond the reach of current biotechnology. But scientists are getting closer with mice, whose embryos have now stayed healthy in vitro for almost half of their 20-day gestation period. Church told me he hopes he'll be manufacturing mice in a lab within five years. And though the elephant's 22-month gestation period is the longest of any mammal, Church said he hopes it will be a short hop from manufacturing mice to manufacturing mammoths.

Church has been thinking about making mammoths for some time, but he accelerated his efforts in 2013, after meeting Sergey Zimov at a de-extinction conference in Washington, D.C. Between sessions, Sergey pitched him on his plan to keep Beringia's permafrost frozen by giving it a top coat of Ice Age grassland. When he explained the mammoth's crucial role in that ecosystem, Church felt compelled to help. He told me he hopes to deliver the first woolly mammoth to Pleistocene Park within a decade.

Last summer, I traveled 72 hours, across 15 time zones, to reach Pleistocene Park. After Moscow, the towns, airports, and planes shrunk with every flight. The last leg flew out of Yakutsk, a gray city in Russia's far east, whose name has, like Siberia's, become shorthand for exile. The small dual-prop plane flew northeast for four hours, carrying about a dozen passengers seated on blue-felt seats with the structural integrity of folding chairs. Most were indigenous people from Northeast Siberia. Some brought goods from warmer climes, including crops that can't grow atop the permafrost. One woman held a bucket of grapes between her knees.

We landed in Cherskiy, a dying gold-mining town that sits on the Kolyma River, a 1,323-mile vein of meltwater, the largest of several that gush out of northeastern Russia and into the East Siberian Sea. Stalin built a string of gulags along the Kolyma and packed them with prisoners, who were made to work in the local mines. Solzhenitsyn called the Kolyma the gulag system's "pole of cold and cruelty." The region retains its geopolitical cachet today, on account of its proximity to the Arctic Ocean's vast undersea oil reserves.

Cherskiy's airstrip is one of the world's most remote. Before it became a Cold War stronghold, it was a jumping-off point for expeditions to the North Pole. You need special government permission to fly into Cherskiy. Our plane had just rolled to a halt on the runway's patchy asphalt when Russian soldiers in fatigues boarded and bounded up to the first row of the cabin, where I was sitting with Grant Slater, an American filmmaker who had come with me to shoot footage of Pleistocene Park. I'd secured the required permission, but Slater was a late addition to the trip, and his paperwork had not come in on time.

Nikita Zimov, who met us at the airport, had foreseen these difficulties. Thanks to his lobbying, the soldiers agreed to let Slater through with only 30 minutes of questioning at the local military base. The soldiers wanted to know whether he had ever been to Syria and, more to the point, whether he was an American spy. "It is good to be a big man in a small town," Nikita told us as we left the base.

Nikita runs the Northeast Science Station, an Arctic research outpost near Cherskiy, which supports a range of science projects along the Kolyma River, including Pleistocene Park. The station and the park are both funded with a mix of grants from the European Union and America's National Science Foundation. Nikita's family makes the 2,500-mile journey from Novosibirsk to the station every May. In the months that follow, they are joined by a rotating group of more than 60 scientists from around the world. When the sky darkens in the fall, the scientists depart, followed by Nikita's family and finally Nikita himself, who hands the keys to a small team of winter rangers.

We arrived at the station just before dinner. It was a modest place, consisting of 11 hacked together structures, a mix of laboratories and houses overlooking a tributary of the Kolyma. Station life revolves around a central building topped by a giant satellite dish that once beamed propaganda to this remote region of the Soviet empire.

I'd barely stepped through the door that first night when Nikita offered me a beer. "Americans love IPAs," he said, handing me a 32-ounce bottle. He led us into the station's dining hall, a warmly lit, cavernous room directly underneath the satellite dish. During dinner, one of the scientists told me that the Northeast Science Station ranks second among Arctic outposts as a place to do research, behind only Toolik Field Station in Alaska. Nikita later confided that he felt quite competitive with Toolik. Being far less remote, the Alaskan station offers scientists considerable amenities, including seamless delivery from Amazon Prime. But Toolik provides no alcohol, so Nikita balances its advantages by stocking his station with Russian beer and crystal-blue bottles of Siberian vodka, shipped into Cherskiy at a heavy cost. The drinks are often consumed late at night in a roomy riverside sauna, under a sky streaked pink by the midnight sun.

Nikita is the life of the station. He is at every meal, and any travel, by land or water, must be coordinated through him. His father is harder to find. One night, I caught Sergey alone in the dining room, having a late dinner. Squat and barrel-chested, he was sitting at a long table, his thick gray rope of a ponytail hanging past his tailbone. His beard was a white Brillo Pad streaked with yellow. He chain-smoked all through the meal, drinking vodka, telling stories, and arguing about Russo-American relations. He kept insisting, loudly and in his limited English, that Donald Trump would be elected president in a few months. (Nikita would later tell me that Sergey has considered himself something of a prophet ever since he predicted the fall of the Soviet Union.) Late in the night, he finally mellowed when he turned to his favorite subjects, the deep past and far future of humankind. Since effectively handing the station over to his son, Sergey seems to have embraced a new role. He has become the station's resident philosopher.

Nikita would probably think *philosopher* too generous. "My dad likes to lie on the sofa and do science while I do all the work," he told me the next day. We were descending into an ice cave in Pleistocene Park. Step by cautious step, we made our way down a pair of rickety ladders that dropped 80 feet through the permafrost to the cave's bottom. Each time our boots found the next rung, we came eye to eye with a more ancient stratum of chilled soil. Even in the Arctic summer, temperatures in the underground network of chambers were below freezing, and the walls were coated with white ice crystals. I felt like we were wandering around in a geode.

Not every wall sparkled with fractals of white frost. Some were windows of clear ice, revealing mud that was 10,000, 20,000, even

30,000 years old. The ancient soil was rich with tiny bone fragments from horses, bison, and mammoths, large animals that would have needed a prolific, cold-resistant food source to survive the Ice Age Arctic. Nikita knelt and scratched at one of the frozen panels with his fingernail. Columns of exhaled steam floated up through the white beam of his headlamp. "See this?" he said. I leaned in, training my lamp on his thumb and forefinger. Between them, he held a thread of vegetable matter so tiny and pale that an errant breath might have reduced it to powder. It was a 30,000-year-old root that had once been attached to a bright-green blade of grass.

For the vast majority of the Earth's 4.5 billion spins around the sun, its exposed, rocky surfaces lay barren. Plants changed that. Born in the seas like us, they knocked against the planet's shores for eons. They army-crawled onto the continents, anchored themselves down, and began testing new body plans, performing, in the process, a series of vast experiments on the Earth's surface. They pushed whole forests of woody stems into the sky to stretch their light-drinking leaves closer to the sun. They learned how to lure pollinators by unfurling perfumed blooms in every color of the rainbow. And nearly 70 million years ago, they began testing a new form that crept out from the shadowy edges of the forest and began spreading a green carpet of solar panel across the Earth.

For tens of millions of years, grasses waged a global land war against forests. According to some scientists, they succeeded by making themselves easy to eat. Unlike other plants, many grasses don't expend energy on poisons, or thorns, or other herbivore-deterring technologies. By allowing themselves to be eaten, they partner with their own grazers to enhance their ecosystem's nutrient flows.

Temperate-zone biomes can't match the lightning-fast bio-cycling of the tropics, where every leaf that falls to the steamy jungle floor is set upon by microbial swarms that dissolve its constituent parts. In a pine forest, a fallen branch might keep its nutrients locked behind bark and needle for years. But grasslands are able to keep nutrients moving relatively quickly, because grasses so easily find their way into the hot, wet stomachs of large herbivores, which are even more microbe-rich than the soil of the tropics. A grazing herbivore returns nutrients to the soil within a day or two, its thick, paste-like dung acting as a fertilizer to help the bitten blades of grass regrow from below. The blades sprout as if from everlasting ribbon dispensers, and they grow faster than any other plant group on Earth. Some bamboo grasses shoot out of the ground at a rate of several feet a day.

Grasses became the base layer for some of the Earth's richest ecosystems. They helped make giants out of the small, burrowing mammals that survived the asteroid that killed off the dinosaurs some 66 million years ago. And they did it in some of the world's driest regions, such as the sunbaked plains of the Serengeti, where more than 1 million wildebeests still roam. Or the northern reaches of Eurasia during the most-severe stretches of the Pleistocene.

The root between Nikita's thumb and forefinger was one foot soldier among trillions that fought in an ecological revolution that human beings would come to join. We descended, after all, from tree-dwellers. Our nearest primate relatives, chimpanzees, bonobos, and gorillas, are still in the forest. Not human beings. We left Africa's woodlands and wandered into the alien ecology of its grassland savannas, as though sensing their raw fertility. Today, our diets—and those of the animals we domesticated—are still dominated by grasses, especially those we have engineered into mutant strains: rice, wheat, corn, and sugarcane.

"Ask any kid 'Where do animals live?' and they will tell you 'The forest,'" Nikita told me. "That's what people think of when they think about nature. They think of birds singing in a forest. They should think of the grassland."

Nikita and I climbed out of the ice cave and headed for the park's grassland. We had to cross a muddy drainage channel that he had bulldozed to empty a nearby lake, so that grass seeds from the park's existing fields could drift on the wind and fall onto the newly revealed soil. Fresh tufts of grass were already erupting out of the mud. Nikita does most of his violent gardening with a forest-mowing transporter on tank treads that stands more than 10 feet tall. He calls it the "mama mammoth."

When I first laid eyes on Pleistocene Park, I wondered whether it was the grassland views that first lured humans out of the woods.

In the treeless plains, an upright biped can see almost into eternity. Cool Arctic winds rushed across the open landscape, fluttering its long ground layer of grasses. On the horizon, I made out a herd of large, gray-and-white animals. Their features came into focus as we hiked closer, especially after one broke into a run. They were horses, like those that sprinted across the plains of Eurasia and the Americas during the Pleistocene, their hooves hammering the ground, compressing the snow so that other grazers could reach cold mouthfuls of grass and survive the winter.

Like America's mustangs, Pleistocene Park's horses come from a line that was once domesticated. But it was hard to imagine these horses being tamed. They moved toward us with a boldness you don't often see in pens and barns. Nikita is not a man who flinches easily, but he backpedaled quickly when the horses feinted in our direction. He stooped and gathered a bouquet of grass and extended it tentatively. The horses snorted at the offer. They stared at us, dignified and curious, the mystery of animal consciousness beaming out from the black sheen of their eyes. At one point, four lined up in profile, like the famous quartet of gray horses painted by torchlight on the ceiling of Chauvet Cave, in France, some 30,000 years ago.

We walked west through the fields, to where a lone bison was grazing. When seen without a herd, a bison loses some of its glamour as a pure symbol of the wild. But even a single hungry specimen is an ecological force to be reckoned with. This one would eat through acres of grass by the time the year was out. In the warmer months, bison expend some of their awesome muscular energy on the destruction of trees. They shoulder into stout trunks, rubbing them raw and exposing them to the elements. It was easy to envision huge herds of these animals clearing the steppes of Eurasia and North America during the Pleistocene. This one had trampled several of the park's saplings, reducing them to broken, leafless nubs. Nikita and I worried that the bison would trample us, too, when, upon hearing us inch closer, he reared up his mighty, horned head, stilled his swishing tail, and stared, as though contemplating a charge.

We stayed low and headed away to higher ground to see a musk ox, a grazer whose entire being, inside and out, seems to have been carved by the Pleistocene. A musk ox's stomach contains exotic microbiota that are corrosive enough to process tundra scrub. Its dense layers of fur provide a buffer that allows it to graze in perfect comfort under the dark, aurora-filled sky of the Arctic winter, untroubled by skin-peeling, 70-below winds.

Nikita wants to bring hordes of musk oxen to Pleistocene Park. He acquired this one on a dicey boat ride hundreds of miles north into the ice-strewn Arctic Ocean. He would have brought back several others, too, but a pair of polar bears made off with them. Admiring the animal's shiny, multicolored coat, I asked Nikita whether he worried about poachers, especially with a depressed mining town nearby. He told me that hunters from Cherskiy routinely hunt moose, reindeer, and bear in the surrounding forests, "but they don't hunt animals in the park."

"Why?" I asked.

"Personal relationships," he said. "When the leader of the local mafia died, I gave the opening remarks at his funeral."

Filling Pleistocene Park with giant herbivores is a difficult task because there are so few left. When modern humans walked out of Africa, some 70,000 years ago, we shared this planet with more than 30 land-mammal species that weighed more than a ton. Of those animals, only elephants, hippos, rhinos, and giraffes remain. These African megafauna may have survived contact with human beings because they evolved alongside us over millions of years—long enough for natural selection to bake in the instincts required to share a habitat with the most dangerous predator nature has yet manufactured.

The giant animals that lived on other continents had no such luxury. When we first wandered into their midst, they may have misjudged us as small, harmless creatures. But by the time humans arrived in southern Europe, we'd figured out how to fan out across grasslands in small, fleet-footed groups. And we were carrying deadly projectiles that could be thrown from beyond the intimate range of an animal's claws or fangs.

Most ecosystems have checks against runaway predation. Population dynamics usually ensure that apex predators are rare. When

Africa's grazing populations dip too low, for instance, lions go hungry and their numbers plummet. The same is true of sharks in the oceans. But when human beings' favorite prey thins out, we can easily switch to plant foods. This omnivorous resilience may explain a mystery that has vexed fossil hunters for more than a century, as they have slowly unearthed evidence of an extraordinary die-off of large animals all over the world, right at the end of the Pleistocene.

Some scientists think that extreme climate change was the culprit: The global melt transformed land-based biomes, and lumbering megafauna were slow to adapt. But this theory has weaknesses. Many of the vanished species had already survived millions of years of fluctuations between cold and warmth. And with a climate-caused extinction event, you'd expect the effects to be distributed across size and phylum. But small animals mostly survived the end of the Pleistocene. The species that died in high numbers were mammals with huge stores of meat in their flanks—precisely the sort you'd expect spear-wielding humans to hunt.

Climate change may have played a supporting role in these extinctions, but as our inventory of fossils has grown, it has strengthened the case for extermination by human rampage. Most telling is the timeline. Between 40,000 and 60,000 years ago, during an ocean lowering glaciation, a small group of humans set out on a sea voyage from Southeast Asia. In only a few thousand years, they skittered across Indonesia and the Philippines, until they reached Papua New Guinea and Australia, where they found giant kangaroos, lizards twice as long as Komodo dragons, and furry, hippo-size wombats that kept their young in huge abdominal pouches. Estimating extinction dates is tricky, but most of these species seem to have vanished shortly thereafter.

It took at least another 20,000 years for human beings to trek over the Bering land bridge to the Americas, and a few thousand more to make it down to the southern tip. The journey seems to have taken the form of an extended hunting spree. Before humans arrived, the Americas were home to mammoths, bear-size beavers, car-size armadillos, giant camels, and a bison species twice as large as those that graze the plains today. The smaller, surviving bison is now the largest living land animal in the Americas, and it barely escaped extermination: The invasion of gun-toting Europeans reduced its numbers from more than 30 million to fewer than 2,000.

The pattern that pairs human arrival with megafaunal extinction is clearest in the far-flung islands that no human visited until relatively recently. The large animals of Hawaii, Madagascar, and New Zealand disappeared during the past 2,000 years, usually within centuries of human arrival. This pattern even extends to ocean ecosystems. As soon as industrial shipbuilding allowed large groups of humans to establish a permanent presence on the seas, we began hunting marine megafauna for meat and lamp oil. Less than a century later, North Atlantic gray whales were gone, along with 95 percent of North Atlantic humpbacks. Not since the asteroid struck have large animals found it so difficult to survive on planet Earth.

In nature, no event happens in isolation. A landscape that loses its giants becomes something else. Nikita and I walked all the way to the edge of Pleistocene Park, to the border between the grassy plains and the forest, where a line of upstart saplings was shooting out of the ground. Trees like these had sprung out of the soils of the Northern Hemisphere for ages, but until recently, many were trampled or snapped in half by the mighty, tusked force of the woolly mammoth.

It was only 3 million years ago that elephants left Africa and swept across southern Eurasia. By the time they crossed the land bridge to the Americas, they'd grown a coat of fur. Some of them would have waded into the shallow passes between islands, using their trunks as snorkels. In the deserts south of Alaska, they would have used those same trunks to make mental scent maps of water resources, which were probably sharper in resolution than a bloodhound's.

The mammoth family assumed new forms in new habitats, growing long fur in northern climes and shrinking to pygmies on Californian islands where food was scarce. But mammoths were always a keystone species on account of their prodigious grazing, their well-digging, and the singular joy they seemed to derive from knocking down trees. A version of this behavior is on display today in South Africa's Kruger National Park, one of the only places on Earth where elephants live in high densities. As the population has recovered, the park's woodlands have thinned, just as they did millions of years ago, when elephants helped engineer the African savannas that made humans into humans.

I have often wondered whether the human who first encountered a mammoth retained some cultural memory of its African cousin, in song or story. In the cave paintings that constitute our clearest glimpse into the prehistoric mind, mammoths loom large. In a single French cave, more than 150 are rendered in black outline, their tusks curving just so. In the midst of the transition from caves to constructed homes, some humans lived *inside* mammoths: 15,000 years ago, early architects built tents from the animals' bones and tusks.

Whatever wonderment human beings felt upon sighting their first mammoth, it was eventually superseded by more-practical concerns. After all, a single cold-preserved carcass could feed a tribe for a few weeks. It took less than 50 millennia for humans to help kill off the mammoths of Eurasia and North America. Most were dead by the end of the Ice Age. A few survived into historical times, on remote Arctic Ocean outposts like St. Paul Island, a lonely dot of land in the center of the Bering Sea where mammoths lived until about 3600 B.C. A final group of survivors slowly wasted away on Wrangel Island, just north of Pleistocene Park. Mammoth genomes tell us they were already inbreeding when the end came, around 2000 B.C. No one knows how the last mammoth died, but we do know that humans made landfall on Wrangel Island around the same time.

The mammoth's extinction may have been our original ecological sin. When humans left Africa 70,000 years ago, the elephant family occupied a range that stretched from that continent's southern tip to within 600 miles of the North Pole. Now elephants are holed up in a few final hiding places, such as Asia's dense forests. Even in Africa, our shared ancestral home, their populations are shrinking, as poachers hunt them with helicopters, GPS, and night-vision goggles. If you were an anthropologist specializing in human ecological relationships, you may well conclude that one of our distinguishing features as a species is an inability to coexist peacefully with elephants.

But nature isn't fixed, least of all human nature. We may yet learn to live alongside elephants, in all their spectacular variety. We may even become a friend to these magnificent animals. Already, we honor them as a symbol of memory, wisdom, and dignity. With luck, we will soon re-extend their range to the Arctic.

"Give me 100 mammoths and come back in a few years," Nikita told me as he stood on the park's edge, staring hard into the fast-growing forest. "You won't recognize this place."

The next morning, I met Sergey Zimov on the dock at the Northeast Science Station. In winter, when Siberia ices over, locals make long-distance treks on the Kolyma's frozen surface, mostly in heavy trucks, but also in the ancestral mode: sleighs pulled by fleet-footed reindeer. (Many far-northern peoples have myths about flying reindeer.) Sergey and I set out by speedboat, snaking our way down from the Arctic Ocean and into the Siberian wilderness.

Wearing desert fatigues and a black beret, Sergey smoked as he drove, burning through a whole pack of unfiltered cigarettes. The twin roars of wind and engine forced him to be even louder and more aphoristic than usual. Every few miles, he would point at the young forests on the shores of the river, lamenting their lack of animals. "This is not wild!" he would shout.

It was early afternoon when we arrived at Duvanny Yar, a massive cliff that runs for six miles along the riverbank. It was like no other cliff I'd ever seen. Rising 100 feet above the shore, it was a concave checkerboard of soggy mud and smooth ice. Trees on its summit were flopping over, their fun-house angles betraying the thaw beneath. Its aura of apocalyptic decay was enhanced by the sulfurous smell seeping out of the melting cliffside. As a long seam of exposed permafrost, Duvanny Yar is a vivid window into the brutal geological reality of climate change.

Many of the world's far-northern landscapes, in Scandinavia, Canada, Alaska, and Siberia, are wilting like Duvanny Yar is. When

Nikita and I had driven through Cherskiy, the local mining town, we'd seen whole houses sinking into mud formed by the big melt. On YouTube, you can watch a researcher stomp his foot on Siberian scrubland, making it ripple like a water bed. The northern reaches of the taiga are dimpled with craters hundreds of feet across, where frozen underground soil has gone slushy and collapsed, causing landslides that have sucked huge stretches of forest into the Earth. The local Yakutians describe one of the larger sinkholes as a "gateway to the underworld."

As the Duvanny Yar cliffside slowly melts into the Kolyma River, it is spilling Ice Age bones onto the riverbank, including woolly-rhino ribs and mammoth tusks worth thousands of dollars. A team of professional ivory hunters had recently picked the shore clean, but for a single 30-inch section of tusk spotted the previous day by a lucky German scientist. He had passed it around the dinner table at the station. Marveling at its smooth surface and surprising heft, I'd felt, for a moment, the instinctive charge of ivory lust, that peculiar human longing that has been so catastrophic for elephants, furry and otherwise. When I joked with Sergey that fresh tusks may soon be strewn across this riverbank, he told me he hoped he would be alive when mammoths return to the park.

The first of the resurrected mammoths will be the loneliest animal on Earth. Elephants are extremely social. When they are removed from normal herd life to a circus or a zoo, some slip into madness. Mothers even turn on their young.

Elephants are matriarchal: Males generally leave the herd in their teens, when they start showing signs of sexual maturity. An elephant's social life begins at birth, when a newborn calf enters the world to the sound of joyous stomping and trumpeting from its sisters, cousins, aunts, and, in some cases, a grandmother.

Mammoth herds were likewise matriarchal, meaning a calf would have received patient instruction from its female elders. It would have learned how to use small sticks to clean dirt from the cracks in its feet, which were so sensitive that they could feel the steps of a distant herd member. It would have learned how to wield a trunk stuffed with more muscles than there are in the entire human body, including those that controlled its built-in water hose. It would have learned how to blast trumpet notes across the plains, striking fear into cave lions, and how to communicate with its fellow herd members in a rich range of rumbling sounds, many inaudible to the human ear.

The older mammoths would have taught the calf how to find ancestral migration paths, how to avoid sinkholes, where to find water. When a herd member died, the youngest mammoth would have watched the others stand vigil, tenderly touching the body of the departed with their trunks before covering it with branches and leaves. No one knows how to re-create this rich mammoth culture, much less how to transmit it to that cosmically bewildered first mammoth.

Or to an entire generation of such mammoths. The Zimovs won't be able to slow the thawing of the permafrost if they have to wait for their furry elephant army to grow organically. That would take too long, given the species's slow breeding pace. George Church, the Harvard geneticist, told me he thinks the mammoth-manufacturing process can be industrialized, complete with synthetic-milk production, to create a seed population that numbers in the tens of thousands. But he didn't say who would pay for it—at the Northeast Science Station, there was open talk of recruiting a science-friendly Silicon Valley billionaire—or how the Zimovs would deploy such a large group of complex social animals that would all be roughly the same age.

Nikita and Sergey seemed entirely unbothered by ethical considerations regarding mammoth cloning or geoengineering. They saw no contradiction between their veneration of "the wild" and their willingness to intervene, radically, in nature. At times they sounded like villains from a Michael Crichton novel. Nikita suggested that such concerns reeked of a particularly American piety. "I grew up in an atheist country," he said. "Playing God doesn't bother me in the least. We are already doing it. Why not do it better?"

Sergey noted that other people want to stop climate change by putting chemicals in the atmosphere or in the ocean, where they could spread in dangerous ways. "All I want to do is bring animals back to the Arctic," he said.

As Sergey and I walked down the riverbank, I kept hearing a cracking sound coming from the cliff. Only after we stopped did I register its source, when I looked up just in time to see a small sheet of ice dislodge from the cliffside. Duvanny Yar was bleeding into the river before our very eyes.

In 1999, Sergey submitted a paper to the journal *Science* arguing that Beringian permafrost contained rich "yedoma" soils left over from Pleistocene grasslands. (In other parts of the Arctic, such as Norway and eastern Canada, there is less carbon in the permafrost; if it thaws, sea levels will rise, but much less greenhouse gas will be released into the atmosphere.) When Beringia's pungent soils are released from their icy prison, microbes devour the organic contents, creating puffs of carbon dioxide. When this process occurs at the bottom of a lake filled with permafrost melt, it creates bubbles of methane that float up to the surface and pop, releasing a gas whose greenhouse effects are an order of magnitude worse than carbon dioxide's. Already more than 1 million of these lakes dot the Arctic, and every year, new ones appear in NASA satellite images, their glimmering surfaces steaming methane into the closed system that is the Earth's atmosphere. If huge herds of megafauna recolonize the Arctic, they too will expel methane, but less than the thawing frost, according to the Zimovs' estimates.

Science initially rejected Sergey's paper about the danger posed by Beringia's warming. But in 2006, an editor from the journal asked Sergey to resubmit his work. It was published in June of that year. Thanks in part to him, we now know that there is more carbon locked in the Arctic permafrost than there is in all the planet's forests and the rest of the atmosphere combined.

For my last day in the Arctic, Nikita had planned a send-off. We were to make a day trip, by car, to Mount Rodinka, on Cherskiy's outskirts. Sergey came along, as did Nikita's daughters and one of the German scientists.

Rodinka is referred to locally as a mountain, though it hardly merits the term. Eons of water and wind have rounded it down to a dark, stubby hill. But in Siberia's flatlands, every hill is a mountain. Halfway up to the summit, we already had a God's-eye view of the surrounding landscape. The sky was lucid blue but for a thin mist that hovered above the Kolyma River, which slithered, through a mix of evergreens and scrub, all the way to the horizon. At the foot of the mountain, the gold-mining town and its airstrip hugged the river. In the dreamy, deep-time atmosphere of Pleistocene Park, it had been easy to forget this modern human world outside the park's borders.

Just before the close of the 19th century, in the pages of this magazine, John Muir praised the expansion of Yellowstone, America's first national park. He wrote of the forests, yes, but also of the grasslands, the "glacier meadows" whose "smooth, silky lawns" pastured "the big Rocky Mountain game animals." Already the park had served "the furred and feathered tribes," he wrote. Many were "in danger of extinction a short time ago," but they "are now increasing in numbers."

Yellowstone's borders have since been expanded even farther. The park is now part of a larger stretch of land cut out from ranches, national forests, wildlife refuges, and even tribal lands. This Greater Yellowstone Ecosystem is 10 times the size of the original park, and it's home to the country's most populous wild-bison herd. There is even talk of extending a wildlife corridor to the north, to provide animals safe passage between a series of wilderness reserves, from Glacier National Park to the Canadian Yukon. But not everyone supports Yellowstone's outward expansion. The park is also home to a growing population of grizzly bears, and they have started showing up in surrounding towns. Wolves were reintroduced in 1995, and they, too, are now thriving. A few have picked off local livestock.

Sergey sees Pleistocene Park as the natural next step beyond Yellowstone in the rewilding of the planet. But if Yellowstone is already meeting resistance as it expands into the larger human world, how will Pleistocene Park fare if it leaves the Kolyma River basin and spreads across Beringia?

The park will need to be stocked with dangerous predators. When they are absent, herbivore herds spread out, or they feel safe enough to stay in the same field, munching away mindlessly until it's over-grazed. Big cats and wolves force groups of grazers into dense, watchful formations that move fast across a landscape, visiting a new patch of vegetation each day in order to mow it with their teeth, fertilize it with their dung, and trample it with their many-hooved plow. Nikita wants to bring in gray wolves, Siberian tigers, or cold-adapted Canadian cougars. If it becomes a trivial challenge to resurrect extinct species, perhaps he could even repopulate Siberia with cave lions and dire wolves. But what will happen when one of these predators wanders onto a city street for the first time?

"This is a part of the world where there is very little agriculture, and very few humans," Sergey told me. He is right that Beringia is sparsely populated, and that continuing urbanization will likely clear still more space by luring rural populations into the cities. But the region, which stretches across Alaska and the Canadian Yukon, won't be empty anytime soon. Fifty years from now, there will still be mafia leaders to appease, not to mention indigenous groups and the governments of three nations, including two that spent much of the last century vying for world domination. America and Russia often cooperate in the interest of science, especially in extreme environments like Antarctica and low-Earth orbit, but the Zimovs will need a peace that persists for generations.

Sergey envisions a series of founding parks, "maybe as many as 10," scattered across Beringia. One would be along the Yentna River, in Alaska, another in the Yukon. A few would be placed to the west of Pleistocene Park, near the Ural mountain range, which separates Siberia from the rest of Russia. As Sergey spoke, he pointed toward each of these places, as if they were just over the horizon and not thousands of miles away.

Sergey's plan relies on the very climate change he ultimately hopes to forestall. "The top layer of permafrost will melt first," he said. "Modern ecosystems will be destroyed entirely. The trees will fall down and wash away, and grasses will begin to appear." The Mammoth Steppe would spread from its starting nodes in each park until they all bled into one another, forming a megapark that spanned the entire region. Humans could visit on bullet trains built on elevated tracks, to avoid disturbing the animals' free movement. Hunting could be allowed in designated areas. Gentler souls could go on Arctic safari tours.

When Sergey was out of earshot, I asked Nikita whether one of his daughters would one day take over Pleistocene Park to see this plan through. We were watching two of them play in an old Soviet-military radar station, about 100 yards from Rodinka's peak.

"I took the girls to the park last week, and I don't think they were too impressed," Nikita told me, laughing. "They thought the horses were unfriendly." I told him that wasn't an answer. "I'm not as selfish as my father," he said. "I won't force them to do this."

Before I left to catch a plane back to civilization, I stood with Sergey on the mountaintop once more, taking in the view. He had slipped into one of his reveries about grasslands full of animals. He seemed to be suffering from a form of solastalgia, a condition described by the philosopher Glenn Albrecht as a kind of existential grief for a vanished landscape, be it a swallowed coast, a field turned to desert, or a bygone geological epoch. He kept returning to the idea that the wild planet had been interrupted midway through its grand experiment, its 4.5-billion-year blending of rock, water, and sunlight. He seems to think that the Earth peaked during the Ice Age, with the grassland ecologies that spawned human beings. He wants to restore the biosphere to that creative summit, so it can run its cosmic experiment forward in time. He wants to know what new wonders will emerge. "Maybe there will be more than one animal with a mind," he told me.

I don't know whether Nikita can make his father's mad vision a material reality. The known challenges are immense, and there are likely many more that he cannot foresee. But in this brave new age when it is humans who make and remake the world, it is a comfort to know that people are trying to summon whole landscapes, Lazarus-like, from the tomb. "Come forth," they are saying to woolly mammoths. Come into this habitat that has been prepared for you. Join the wolves and the reindeer and the bison who survived you. Slip into your old Ice Age ecology. Wander free in this wild stretch of the Earth. Your kind will grow stronger as the centuries pass. This place will overflow with life once again. Our original sin will be wiped clean. And if, in doing all this, we can save our planet and ourselves, that will be the stuff of a new mythology.

The Eaten World

By Nitin K. Ahuja

(Originally appeared in
the *New Inquiry*,
December 2016)

A few months ago I had my gut microbiome analyzed–a $99 value–free of charge. The American Gastroenterology Association made the offer to all of its trainee members, myself included. Through the mail I received two cotton swabs in a sterile plastic tube (the accompanying instructions were careful to specify that the amount of stool needed for a successful analysis was likely less than I expected). I sent back the swabs, no longer sterile, and a few weeks later looked up my results online.

As a gastroenterology fellow, I hear a fair bit about the microbiome–a term referring to the aggregate genetic material of the microorganisms that colonize our bodies' various surfaces–as a vast new frontier for biomedicine. In general, I try to maintain some skepticism about the whole thing. The science is exciting but young: studies of gut flora tend to yield associative rather than causative conclusions, and the results of most animal studies are often hard to reproduce in humans. Experimental techniques vary by laboratory, and the best methods for storing and culturing samples are matters of ongoing debate. Conventionally undetectable bacterial species, so-called "microbial dark matter," and innumerable ambient viruses in the gastrointestinal tract constitute a large set of potentially confounding variables.

All the same, when my personalized readout became available, I scrutinized it at length, working to discern my bill of health from a smear of stool at a single point in time. There was a bar graph comparing my dominant bacterial subclasses with those of other participants matched to me by age and gender. Heat maps and scatter plots triangulated the composition of my microbiome with that of other Americans and, by way of cultural contrast, a cohort of Malawians. My sample, I was told, contained at least ten rare taxa, the names of which pleased me, though I had never heard them before–*Dorea, Dermabacter, Anaerofustis*.

Funhouse mirrors are still mirrors, prone to catching the eye. Knowing full well the limits of these metrics, I found myself reading through the latest journal articles for a sense of what the results might mean for me and my body–functional for now, but charged over the last few years by every passing ache or pain with a sense of its pending brokenness.

In theorizing about our health, humans have a longstanding habit of mapping our bodies to the wider world. This pattern runs through a variety of healing systems as a string of what certain historians of medicine have called "microcosm/macrocosm homologies"–explanatory parallels between small and large spaces. Physical regulation is compared with the political state in ancient Chinese writing, for instance, and with the cosmos in Ayurvedic texts. In 16th century Europe, the scholar Paracelsus attributed anatomically specific illnesses to toxins emitted by celestial forms–the spleen was poisoned by Saturn, the liver by Jupiter, the brain by the moon. He further postulated that the affected organs could be restored to health by plants that bore them visual resemblance.

The symbolic underpinnings of the modern microbiome concept are similarly arranged. The gut, in particular, is host to a collection of organisms commensurate in number and complexity with a world in itself; and recent studies have accorded these organisms with progressive influence over our physical and mental wellbeing. The resulting biological model figures the human body as an intermediary between the microcosm it contains and the macrocosm it inhabits.

Both realms are subject to similar threats–a loss of biodiversity attributable to our own behavior. Doubt has fallen over a range of contemporary practices shown to disrupt the composition of our microbiota, ranging from the overuse of antibiotics, to the frequency of C-sections, to the low-fiber, high-fat diet shorthanded as "Westernized." Intuitively, these practices seem to correlate with much more visible ravages of civilization on the global landscape–longstanding environmental challenges like pollution and climate change, the worsening of which appears to be increasingly and grimly certain.

This explicit mapping between inner and outer worlds has helped establish new priorities within the modern biomedical mission. Public health advocates like Martin Blaser, a pioneering microbiologist and former chair of medicine at NYU, often utilize this analogy in calls for a more scrupulous approach to clinical intervention. Blaser's 2014 book, *Missing Microbes*, connects different types of environmental instabilities to build the case for a novel understanding of sickness and health. "I see many parallels between our changing climate and our changing resident microbes," he writes, speculating on the increasing prevalence of diseases like asthma and obesity. "Like the worsening hurricane seasons we are seeing, these outcomes are bad enough, but they are also indicators of our larger imbalances, the loss of our reserves."

This notion of balance is a longstanding feature of medical thinking, from classical formulations of humoral theory to contemporary models of biochemical homeostasis. As such, an understanding of health as a consequence of maintaining the right parts in the right proportions—whether phlegm and bile or Firmicutes and Bacteroidetes—comes with deep-seated narrative appeal. And yet, the mechanism of disease implied by the microbiome model is also one particularly well suited to the contemporary moment. It provides a new mode of expressing an old distress, transforming the unease with which we regard our effects on the natural world into a literally visceral phenomenon.

Balance is likewise a core metaphor for the modern environmental movement, latent in the conceptual understanding of ecosystems and sustainability. Analogous threats breed analogous anxieties, in light of which our contemporary fascination with the microbiome seems distinctly Anthropocenic—belonging to this putative geological age defined by human activity. We've taken the world, as we've lately come to see it, into ourselves—depleted, worn away, eaten up.

The ecological analogy is often justified. Microbiome research consists of explicitly transdisciplinary questions, and the importation of tools from the environmental sciences is to some extent obligatory. Like any

"We've taken the world, as we've lately come to see it, into ourselves—depleted, worn away, eaten up."

complicated idea, though, the microbiome is defined by its boldest outlines when distilled for popular consumption.

How might this framing fail us? To the extent that descriptive parallels set the stage for moral ones, the gaps left by science are well positioned to be filled by sentiment. Mapping our bodies to an endangered world has a way of skewing the biomedical dialogue toward stock ecological narratives in which the trappings of civilization are inherently destructive, or the most valuable landscape is the one least shaped by human hands.

In this regard, the reparative spirit that has emerged in the public sphere with respect to the microbiome seems pre-loaded with a sense of culpability. Recently published self-help books on gut health favor a rhetoric of loss and rehabilitation in their service as guides to, for example, "restore your inner ecology." Coursing beneath this nostalgia for times past is a subtle sort of primitivism, the knee-jerk tendency to view indigenous cultures as time capsules for physical vigor that civilization has elsewhere degraded. Data suggest that microbial diversity is often higher in remote populations, but the health implications of this difference remain conjectural. Even so, the idea was credible enough to inspire at least one independent researcher to introduce the stool of a Tanzanian hunter-gatherer into his own rectum with a turkey baster.

A wounded ecological reading of the microbiome is not inevitable. Science journalist Ed Yong's book, *I Contain Multitudes*, published this past August, is an overview for general audiences that favors a tone of wonder over warning. It does so in large part by emphasizing the insignificance of humanity relative to bacteria in evolutionary history. For example, in response to the idea of the Anthropocene, Yong writes, "You could equally argue that we are still living in the Microbiocene: a period that started at the dawn of life itself and will continue to its very end." While acknowledging our real capacity to damage microbial ecosystems beyond repair, the author humbles himself repeatedly to those ecosystems' complexity and resilience. Whereas Blaser's analysis functions as an alarm call, Yong's is rendered as a celebration of biological peculiarity, all that we don't yet know.

Macro-environmental theory, for its part, has matured beyond the tidy concept of unspoiled wilderness. Contemporary ecocritics argue that stasis is a false ideal, deployed primarily for aesthetic effect; it represents a contrivance of "pastoral ecology," a quaint and outmoded view of nature typified by a harmonious resting state. As such, distinctions between human and non-human worlds may be viewed as increasingly arbitrary with respect to the primary goal of safeguarding both. Peter Kareiva, former chief scientist at The Nature Conservancy, a well-known charitable organization, has provocatively suggested that modern environmentalism would do well to relinquish the old metaphors of a fallen world and embrace a more durable understanding of nature. Granting the world some buoyancy in the face of human enterprise, he maintains, will allow us to develop more creative, inclusive, and potent solutions as its stewards moving forward.

Recognizing that this stance is controversial with respect to the planet's shared spaces, it seems considerably less problematic when applied to the space of a single colon. Without diminishing in any way the urgency of the larger conservationist movement or the startling irresponsibility of its dismissal, it should be noted that there are differences between these two types of environments. Unlike rainforests or coral reefs, my body is just over thirty years old and is mine to steward alone. Microbes connect us vertically across generations and horizontally with every handshake, but the world inside me–however balanced, diverse, or protected–is fated to collapse whenever I do.

Individual profiles of the microbiome may yet mature into meaningful portraits. Despite the continued relevance of an eye-rolling award regularly given out by evolutionary biologist Jonathan Eisen for "Overselling the Microbiome," the field tends to be savvy to its own limitations. Recognizing that bacterial diversity is a moving target, researchers have become more attentive to obtaining multiple samples over time, despite the extra costs of this approach. Interest in the relative abundance of species is coupled with progressive interest in their associated metabolome, the collection of biologically active molecules that those species produce. In general, methods such as these work toward establishing a more nuanced understanding of the relationship between microorganisms and human health.

At the same time, the unfinished quality of microbiome science has not precluded attempts to commodify it. Start-ups built around microbial therapies have surged in value on the basis of hype, and some have already crashed upon failure to deliver. The scale and enthusiasm of these ventures seem driven as much by the scientific potential of the microbiome as by the affective power of its conceptual model. There is a clear imperative to invest, both financially and emotionally, in the future of this microcosm.

For now, publicly available microbial analyses are like Rorschach blots, reliant on the viewer for active interpretation. The tone of many readings tends toward penitence: in parallel with ecological convention, we are poised to see our civilized guts as reflective of descent from an earlier, healthier, more natural state of being. So far, these little landscapes have struck us as more fragile than plastic, but the basis for this outlook at a planetary scale does not necessarily hold firm at an intestinal one. With respect to the microbiome, recognizing the sentimental parameters of this particular homology can help enlarge our therapeutic vision and bring it, at least partially, under our control.

I've been asked by patients with all sorts of symptoms if they might be considered for a stool transplant—the instillation of fecal microbiota from a healthy donor into the bowels of an unwell recipient (but with a colonoscope instead of a turkey baster). The formal diagnoses for which this procedure has been studied and approved are in fact quite limited. In some corners of the public consciousness, though, the idea of resetting one's intestinal bacteria has taken on the dimensions of panacea. Prelapsarian ideals of personal and environmental health are perhaps subject to similar charges of romanticism and naiveté, and I readily confess to finding them quite charming myself. The impulse they draw upon, after all, is reflexive and tenderhearted—to replant the garden, to start over again.

A Forest of Furniture Is Growing In England

By Sarah Laskow

(Originally appeared in
Atlas Obscura,
December 2017)

T he rain was streaming down on the summer afternoon in 2012 when Gavin Munro realized with chagrin that he had gotten exactly what he'd wished for.

He had spent the morning hunched over in his black slicker, working willow branches into new shapes. Lunch had been miso soup and oatcakes, same as every day, and after that frugal meal it had been hard to convince himself to return to the field. It was 3:00, when he paused each afternoon to take a picture, as a marker of time in a decade-long plan. Even now, standing in a forest of chairs sprouting directly from the ground, it was hard to believe that plan could work.

He paused a long moment, alone, before stooping back down, thinking—"Shit, I got what I asked for. I'm a chair farmer."

Six years earlier, in 2006, Munro had started work on his grand vision. It was a big, mad idea that he had become so obsessed with that one friend had picked a safe word—"haberdashery"—that would make him stop talking about it. Munro wanted to change the way people think about manufacturing. It might take a day to assemble enough flat-pack furniture to fill a house, but the timber cut down to make it all needs decades to grow. Even the cheapest wooden chairs require a wealth of time to create. Munro's big idea was that he would guide trees to grow into chairs, tables, and lamps that could be harvested right out of a field. The trees, selected for their ability to grow new sprouts from their stumps, would regenerate. His forest would yield furniture the way an orchard yields apples.

He started a company, Full Grown, to pursue his plan, and that day in 2012 when he stopped and took stock of where it had led him, he was more than halfway through the decade that he had given himself to grow his first crop. He had yet to harvest a single chair.

The narrow road to Munro's chair forest runs between close stone walls, along a route worn deep by years of cattle walking into Wirksworth, England. Ed Lound, who said he'd be easy to recognize because no one else in town has blond dreadlocks, knows exactly which bits of road can accommodate both his jeep and a car heading in the other direction. His family moved here when he was five years old, which makes him a newcomer in what he calls a "really old place." Some local historians believe Wirksworth is the site of the lost Roman city of Lutudarum, which had been built to mine lead from somewhere here in Derbyshire. In the parish church, there's a piece of old stonework, originally found in a 13th-century building, that shows a man holding a pick and a basket. It's said to be the oldest representation of a miner anywhere in the world: T'owd Man.

People who've spent most of their lives in this town—even those who were born here but whose families came from elsewhere—still aren't considered true Wirksworthians. Munro, born one town over, in Matlock, is as much a foreigner as the wealthy interlopers from the south buying up local real estate. If the roots that anchor Munro and Lound to this place are shallower than others', they're still part of a deep, strong network. Munro, whose wife grew up in Wirksworth, met Lound through friends and hired him in 2014, fresh out of the University of York with a degree in criminology. More than three years later, Lound knows the 2,000 or so trees in the forest as well as Munro does.

Just inside a gate, there are rows of ash, oak, sycamore, and hazel, alongside red-headed beech and self-seeding goat willow. Each individual tree is being formed into a piece of furniture. Saying that you're going to visit a forest of chairs in the English Midlands sounds like you're embarking on an adventure in Narnia, and in Munro's original vision he imagined chairs and tables lined up in neat orchard rows. The field is wilder than that. Berries grow among the trees, pheasants and rabbits nestle in the grass, harvest mice live in the shed, and birds nest in the lamps. But it doesn't feel so different from other agricultural places until you walk down rows and see trees with, simultaneously, all their usual attributes—branches, leaves, roots— and all the attributes of chairs—backs, seats, legs, set at right angles.

The chairs grow upside-down, their four legs stretching up toward the sky. Lound grabs hold of one that's almost ready for harvest. "It's thickening up at the right level," he says, as if describing a prized farm animal. "It's just level and sturdy. If you do that"—he shakes the branch—"the whole tree moves."

We're looking at one of the most promising chairs in the field, which represents years of trial and error. According to Munro's original plan, the first crop of chairs should have been harvested by 2016, but most of the pieces, more than 500 in all, are still in the field, including a row of squat, spiral lamps planned as a quick cash crop. "Making trees do what they don't want to do is really bad, and see how shallow we've laid these branches?" says Lound, pointing at one of the lamps. "That's not what a tree wants to do."

A tree has a basic plan, embedded in its cells, for optimizing its position in the world. When Munro first started to experiment with training trees, he tried to join sets of four, with one tree for each leg, into chairs that were supposed to grow from the ground up. The trees resisted. Planted close together, they competed for light and space, and one would always come to dominate at the expense of its neighbors. Munro was bossy with them, too, and would nudge them in directions they wouldn't usually go. They grew cautious. Their progress slowed, or they gave up on branches he had bent into his designs, which left some chair-parts too skinny and underdeveloped, while others continued to fatten.

For every branch that becomes part of the chair, the tree wants to grow many more. Guiding a tree's growth begins with selecting branches that seem most naturally inclined to reach in a given direction. At the beginning, the tree doesn't look much like a tree at all, but a wide T-shape stuck into the earth. As more branches bud and grow, the most amenable ones are tied to frames that keep them in line. Later they're bent to form the angles of the chair's seat or legs. Pruning slows the growth of parts of the chair while other parts continue to develop, and grafts join branches together to make the chairs' legs.

Full Grown's earlier chair designs required accommodations from the tree: One branch, for instance, was supposed to split into two, with

"All along, Munro knew it should be possible to coax a tree into becoming a chair."

one limb continuing skyward as part of a back leg while the other became part of the chair's seat and then was curved to form a front leg. But the tree favored the back leg, so the design had to change, with separate branches forming each leg. "We were trying to design it like a chair," says Lound. "Whereas now we're trying to grow a tree."

All along, Munro knew it should be possible to coax a tree into becoming a chair. In the Full Grown office he keeps a picture of the throne-like "Chair That Grew," started from seed and harvested in 1914 by John Krubsack, a Wisconsin banker and gentleman farmer. In the century since, others have independently rediscovered the idea of tree-shaping, always working alone, rarely sharing their knowledge with anyone else.

Among the few people in the world who have dedicated themselves to this craft, bending trees to their own designs, Axel Erlandson, a farmer in California, was the first to turn it into an art form. "He was really the master," says Lound. "He got it down, and then he died without telling anyone how he did it, which really didn't help at all."

Erlandson was dedicated to precision, able to convince his trees to grow exactly as he imagined they could. He had worked as a surveyor, and he drafted plans for his trees as if he was creating a map, accurate to 1/1000 of a foot. In 1929, four years after he began this work, he sketched out a design for his Poplar Archway, which required 10 trees, planted 18 inches apart, to grow into a lattice of Gothic-like windows, with a three-foot doorway set in the center. His wife was skeptical. She wrote, on the same page as his drawing, "I do not believe that Axel can get a tree to grow like this illustration." On the other half of the page, he wrote, "I believe I can get a tree to grow like this illustration." He was correct.

Erlandson's formal schooling had stopped after fourth grade, but he had always been adept at understanding complicated machines. He had built a working wooden model of a threshing machine as a young man in Minnesota and, later, windmills to suck up water and irrigate his arid California farmland. He owned a motorcycle, in days when motorcycles needed even more constant attention than they do today, and drove it across the country. To him, the workings of trees posed a new puzzle.

The idea of training trees into unique shapes came to him on his farm in the Central Valley. He had planted a row of trees as a windbreak for his crops, and he noticed that some branches lost their bark and began to graft together as the wind rubbed them against one another. Experimenting with the shape of trees became a hobby. One of the first trees he designed, the Four-Legged Giant, was a sycamore made of four individual trees grafted together to become one, which straddled the ground like an invader from Mars.

He created wonderful, fantastical shapes, which no one had dreamed trees could form. There was the Two-Legged Tree, which stood astride a pathway with a pair of perfectly arched legs, and trees with trunks that branched into circles, cubes, and spheres, before rejoining and growing straight up toward the sky. There was a tree grown into a double-helix, a tree ladder, a base turned into a cage that a person could step inside. Another creation, the Basket Tree, had a towering tube of diamond latticework in place of a trunk.

How he shaped these trees with such precision is a mystery. He believed that he had only begun to discover the possibilities of this art, and that "a person could grow a grove of trees of designs so much more intricate than I have here that they would make my present place look quite simple in comparison," he wrote in a letter in 1953. But he never taught anyone his craft before he died, nine years later.

"I think he enjoyed that people were just amazed," says Mark Primack, a Santa Cruz architect whose fascination with living structures led him to Erlandson's trees. Primack is responsible for collecting many of the details of Erlandson's life that are known today, and helped save the inhabitants of the Tree Circus when the property was in danger of development. "He may have done grafts and stuck nails through the branches to hold them together, but that's all covered up now and internalized," Primack says. "As one of his trees ages, it grows over all his ministrations. It just becomes a tree, a form. There are no signs of how he did it."

Erlandson considered his work, in the idiom of World War II, a "non-essential industry." In 1946, his wife and daughter convinced him to start a "Tree Circus" in Santa Cruz, where they charged visitors 25 cents to walk between the legs of the Four-Legged Giant and into a park filled with the "world's strangest trees." But it failed to thrive. Business was never strong, and when a new highway bypassed the Circus' road, even fewer customers found their way there.

"The principle things we need this world are surely food, clothing, and shelter, and growing these kind of trees can hardly help to satisfy any of those needs," Erlandson wrote in that 1953 letter.

But there are people now who believe that techniques like his can be used to serve more practical ends. A group of German architects, engineers, and scientists are developing principles of experimental "Baubotanik" architecture, in which trees are grown into shelters. One of the test projects is a three-story tower in a lattice pattern reminiscent of Erlandson's work. And at Full Grown, Munro and his colleagues believe that they can create a woodland that regularly produces furniture—a "factory where birds will like living," as Munro puts it. Far into the future, people still are going to want places to sit and surfaces to eat from, and if the company does create a new mode of mass production, a forest that grows chairs, it could provide durable, beautiful objects to serve basic human needs.

The strength of Full Grown's furniture comes, in part, from the grafts that knit the branches together. When the top layer of bark is rubbed away, leaving a layer of growth cells exposed, that wood will join with the tissue of another branch, so that the two limbs grow together as if they were one. A friend of Munro's once pointed out that he must know what that feels like, in a way. Munro was born with vertebrae in his neck fused and malformed, a rare bone disease called Klippel-Feil syndrome, and when he was young, doctors broke those bones and reset them. He had to wear a metal frame that kept his spine in place while the bone grafts healed. Even today, long hours in the field can leave him in pain.

When Munro first started growing trees in the field, he used to huddle for warmth where the cows from the next field over would congregate. Eventually he pulled a small RV onto the site, and later upgraded to a more permanent shed. For the past two winters, though, the company has had an office in town, in a low building that was once part of an engine-block facility. Now the warmth comes from a space heater and tea instead of a herd of cows.

On the office wall, Munro has laid out a new timeline—this one made of bright sticky notes—that imagines his company's growth years into the future. "We're trying to make the transition from art students and a criminology student into actually running a business," says Munro. They've been taking preorders for chairs, tables, and lamps, but they eventually want to be able to sell off-the-shelf, finished pieces. The timeline goes out to 2031 and includes Lound's 30th birthday, as well as the 50th for Munro and Chris Robinson, who went to art school with Munro and is now helping in the field and with new designs. "By the time Ed's 30 we're going to be sitting around a giant dining table that we grew. By the time Chris is 50, we're going to be planting a new farm on land we own," Munro says. "By the time I'm 50, everything will be smooth sailing."

In the chilly back room, a small collection of furniture that has been harvested from the field is drying out. Curlicue lamps hang from the ceiling, and leggy table prototypes crowd together in a corner. Some pieces are still raw, covered in bark, but a few have had some of their surfaces sanded down to smooth, blond wood. In this room is also a single chair, harvested this past fall. Since 2012, other prototypes have come out of the field, but this is the first Full Grown chair that it's possible to sit on with confidence. Munro has been using it, on occasion, as an office chair.

The chair, shorn of its leafy branches and turned right-side up, looks unashamedly chair-like, apart from the nubbin on the curve of its back, where it was once connected to a trunk. It's still covered in bark, though, and retains an essential tree-ness. The legs are sturdy, and where most chairs have a flat slab to sit on, this one has set of curved branches. However, one side of the seat's front edge is too fragile: The tree had directed its resources elsewhere. "But you can lean back, and it's quite comfy," Munro says, demonstrating the gentlest way to sit down. It's taken more than a decade to reach this simple moment.

The chair is strong and supportive. The branches that make the back have a slight give, and the edges wrap ever so slightly around the shoulders of a slim person. Leaning back feels like a gentle embrace, like a trust fall with a tree.

Before harvesting any more chairs, Munro wants to make sure their branches grow even thicker than these, so they leave smaller gaps in the seat and back. After they're harvested and dried, their outer edges will be planed and sanded silky smooth. Any surface that might touch a person's body will get this treatment, but the undersides of the arms and seat, along with other interior surfaces, will keep their bark. The exposed wood's then rubbed down with oil for polish and smoothness.

It won't look exactly like any other chair, but ideally it won't be immediately obvious how the chair was made, either. "Just so that it's not as in your face," says Robinson. "So it's not the first thing you think of. The first thing you think is, 'Oh, that's a really nice chair.'"

Munro goes to the bookshelf and comes back with a tome of Euclidean geometry, full of bright lines in primary colors, that describes basic shapes and their rules. One of his close-held desires is to grow a piece in the shape of a cube, and he wants Full Grown's furniture to have geometric, midcentury qualities, in contrast with the natural texture of the wood and the bark left in place. "We're just applying these rules in four dimensions," says Robinson. "We're involving time as well."

The chair itself reveals the time that went into it. The piece of trunk left on the back reveals its age rings, the story of the years it spent in a field in Derbyshire. Emphasizing that time adds to the chair's aura: "There's that same quality that you get with wines and whisky, of age and time, and there's no substitute for that," says Munro.

In the 11 years since he started this work, Munro, along with Lound, Robinson, and a small band of other recruits, has learned how to work with a tree's natural inclinations, just as Erlandson once did. Now they know to spiral a lamp's branches upwards at 45 degree angles, a more natural path for tree growth. Twisted into this shape, a tree will grow about as much in one year as it would have in five years struggling in their older design. The newer chair designs, too, are dictated by the tree's preferences and require fewer interventions, accommodations, and compromises.

"The tree doesn't necessarily want to grow into a chair. But at the same time we're discovering that we can make growing into the chair quite comfortable and reasonable. And actually quite nice for the tree," says Munro.

The most dramatic intervention they make in the process is sawing a fully grown piece of furniture off its trunk. But even reduced to a stump, as long as part of the trunk survives, these trees will put out new branches, fast and plentiful, to match their strong masses of roots below. It's a technique passed down over thousands of years of sustainable wood-harvesting in this part of the world. "If we cut the tree down and it just died, that's pointless," says Munro. Instead, after the first harvest, the trees grow back quicker and stronger.

Closing in on the goals of his first 10-year plan, it seems possible to Munro that his vision, of growing everyday objects from trees, could become reality. "If we want objects, and we do, then we could quietly make them like this. Every time we reach a parameter, we can figure out a way around it," he says. "We haven't gotten to the edge of it yet. I think Ed might be an old man before we get to the edge of it."

Munro has another timeline in his head, too, this one more speculative. This is for after he is dead, Chris is dead, Ed is dead. It goes out to 5017, as far as his imagination can take him. On that faraway horizon, these techniques are being used to grow more than just chairs and lamps; they've helped reimagine the ways we make objects altogether. This might seem like a distant future, but, to the extent that we depend on trees, we already interact with living things that know this time scale. The oldest tree in the world, a bristlecone pine in the California mountains, is 5,000 years old. To reshape the world means believing in the promise of sweeping, far-sighted plans, and understanding that two millennia might only be half a lifetime. It means thinking in tree time.

"London Bridge is Down": The Secret Plan for the Days After the Queen's Death

By Sam Knight

(First appeared in the *Guardian*, March 2017)

I n the plans that exist for the death of the Queen – and there are many versions, held by Buckingham Palace, the government and the BBC – most envisage that she will die after a short illness. Her family and doctors will be there. When the Queen Mother passed away on the afternoon of Easter Saturday, in 2002, at the Royal Lodge in Windsor, she had time to telephone friends to say goodbye, and to give away some of her horses. In these last hours, the Queen's senior doctor, a gastroenterologist named Professor Huw Thomas, will be in charge. He will look after his patient, control access to her room and consider what information should be made public. The bond between sovereign and subjects is a strange and mostly unknowable thing. A nation's life becomes a person's, and then the string must break.

There will be bulletins from the palace – not many, but enough. "The Queen is suffering from great physical prostration, accompanied by symptoms which cause much anxiety," announced Sir James Reid, Queen Victoria's physician, two days before her death in 1901. "The King's life is moving peacefully towards its close," was the final notice issued by George V's doctor, Lord Dawson, at 9.30pm on the night of 20 January 1936. Not long afterwards, Dawson injected the king with 750mg of morphine and a gram of cocaine – enough to kill him twice over in order to ease the monarch's suffering, and to have him expire in time for the printing presses of the Times, which rolled at midnight.

Her eyes will be closed and Charles will be king. His siblings will kiss his hands. The first official to deal with the news will be Sir Christopher Geidt, the Queen's private secretary, a former diplomat who was given a second knighthood in 2014, in part for planning her succession.

Geidt will contact the prime minister. The last time a British monarch died, 65 years ago, the demise of George VI was conveyed in a code word, "Hyde Park Corner", to Buckingham Palace, to prevent switchboard operators from finding out. For Elizabeth II, the plan for what happens next is known as "London Bridge." The prime minister will be woken, if she is not already awake, and civil servants will say "London Bridge is down" on secure lines. From the Foreign Office's Global Response Centre, at an undisclosed location in the capital, the news will go out to the 15 governments outside the UK where the Queen is also the head of state, and the 36 other nations of the Commonwealth for whom she has served as a symbolic figurehead – a face familiar in dreams and the untidy drawings of a billion schoolchildren – since the dawn of the atomic age.

For a time, she will be gone without our knowing it. The information will travel like the compressional wave ahead of an earthquake, detectable only by special equipment. Governors general, ambassadors and prime ministers will learn first. Cupboards will be opened in search of black armbands, three-and-a-quarter inches wide, to be worn on the left arm.

The rest of us will find out more quickly than before. On 6 February 1952, George VI was found by his valet at Sandringham at 7.30am. The BBC did not broadcast the news until 11.15am, almost four hours later. When Princess Diana died at 4am local time at the Pitié-Salpêtrière hospital in Paris on 31 August 1997, journalists accompanying the former foreign secretary, Robin Cook, on a visit to the Philippines knew within 15 minutes. For many years the BBC was told about royal deaths first, but its monopoly on broadcasting to the empire has gone now. When the Queen dies, the announcement will go out as a newsflash to the Press Association and the rest of the world's media simultaneously. At the same instant, a footman in mourning clothes will emerge from a door at Buckingham Palace, cross the dull pink gravel and pin a black-edged notice to the gates. While he does this, the palace website will be transformed into a sombre, single page, showing the same text on a dark background.

Screens will glow. There will be tweets. At the BBC, the "radio alert transmission system" (Rats), will be activated – a cold war-era alarm designed to withstand an attack on the nation's infrastructure. Rats, which is also sometimes referred to as "royal about to snuff it", is a near mythical part of the intricate architecture of ritual and rehearsals for the death of major royal personalities that the BBC has maintained since the 1930s. Most staff have only ever seen it work in tests; many have never seen it work at all. "Whenever there is a strange noise in the newsroom, someone always asks, 'Is that the Rats?' Because we don't know what it sounds like," one regional reporter told me.

All news organisations will scramble to get films on air and obituaries online. At the Guardian, the deputy editor has a list of prepared stories pinned to his wall. The Times is said to have 11 days of coverage ready to go. At Sky News and ITN, which for years rehearsed the death of the Queen substituting the name "Mrs Robinson", calls will go out to royal experts who have already signed contracts to speak exclusively on those channels. "I am going to be sitting outside the doors of the Abbey on a hugely enlarged trestle table commentating to 300 million Americans about this," one told me.

For people stuck in traffic, or with Heart FM on in the background, there will only be the subtlest of indications, at first, that something is going on. Britain's commercial radio stations have a network of blue "obit lights", which is tested once a week and supposed to light up in the event of a national catastrophe. When the news breaks, these lights will start flashing, to alert DJs to switch to the news in the next few minutes and to play inoffensive music in the meantime. Every station, down to hospital radio, has prepared music lists made up of "Mood 2" (sad) or "Mood 1" (saddest) songs to reach for in times of sudden mourning. "If you ever hear Haunted Dancehall (Nursery Remix) by Sabres of Paradise on daytime Radio 1, turn the TV on," wrote Chris Price, a BBC radio producer, for the Huffington Post in 2011. "Something terrible has just happened."

Having plans in place for the death of leading royals is a practice that makes some journalists uncomfortable. "There is one story which is deemed to be so much more important than others," one former Today program producer complained to me. For 30 years, BBC news teams were hauled to work on quiet Sunday mornings to perform mock storylines about the Queen Mother choking on a fishbone. There was once a scenario about Princess Diana dying in a car crash on the M4.

These well-laid plans have not always helped. In 2002, when the Queen Mother died, the obit lights didn't come on because someone failed to push the button down properly. On the BBC, Peter Sissons, the veteran anchor, was criticized for wearing a maroon tie. Sissons was the victim of a BBC policy change, issued after the September 11 attacks, to moderate its coverage and reduce the number of "category one" royals eligible for the full obituary procedure. The last words in Sissons's ear before going on air were: "Don't go overboard. She's a very old woman who had to go some time."

But there will be no extemporizing with the Queen. The newsreaders will wear black suits and black ties. Category one was made for her. Programs will stop. Networks will merge. BBC 1, 2 and 4 will be interrupted and revert silently to their respective idents – an exercise class in a village hall, a swan waiting on a pond – before coming together for the news. Listeners to Radio 4 and Radio 5 live will hear a specific formulation of words, "This is the BBC from London," which, intentionally or not, will summon a spirit of national emergency.

The main reason for rehearsals is to have words that are roughly approximate to the moment. "It is with the greatest sorrow that we make the following announcement," said John Snagge, the BBC presenter who informed the world of the death of George VI. (The news was repeated seven times, every 15 minutes, and then the BBC went silent for five hours). According to one former head of BBC news, a very similar set of words will be used for the Queen. The rehearsals for her are different to the other members of the family, he explained. People become upset, and contemplate the unthinkable oddness of her absence. "She is the only monarch that most of us have ever known," he said. The royal standard will appear on the screen. The national anthem will play. You will remember where you were.

When people think of a contemporary royal death in Britain, they think, inescapably, of Diana. The passing of the Queen will be monumental by comparison. It may not be as nakedly emotional, but its reach will be wider, and its implications more dramatic. "It will be quite fundamental," as one former courtier told me.

Part of the effect will come from the overwhelming weight of things happening. The routine for modern royal funerals is more or less familiar (Diana's was based on "Tay Bridge", the plan for the Queen Mother's). But the death of a British monarch, and the accession of a new head of state, is a ritual that is passing out of living memory: three of the Queen's last four prime ministers were born after she came to the throne. When she dies, both houses of parliament will be recalled, people will go home from work early, and aircraft pilots will announce the news to their passengers. In the nine days that follow (in London Bridge planning documents, these are known as "D-day", "D+1" and so on) there will be ritual proclamations, a four-nation tour by the new king, bowdlerised television programming, and a diplomatic assembling in London not seen since the death of Winston Churchill in 1965.

More overwhelming than any of this, though, there will be an almighty psychological reckoning for the kingdom that she leaves behind. The Queen is Britain's last living link with our former greatness – the nation's id, its problematic self-regard – which is still defined by our victory in the second world war. One leading historian, who like most people I interviewed for this article declined to be named, stressed that the farewell for this country's longest-serving monarch will be magnificent. "Oh, she will get everything," he said. "We were all told that the funeral of Churchill was the requiem for Britain as a great power. But actually it will really be over when she goes."

Unlike the US presidency, say, monarchies allow huge passages of time – a century, in some cases – to become entwined with an individual. The second Elizabethan age is likely to be remembered as a reign of uninterrupted national decline, and even, if she lives long enough and Scotland departs the union, as one of disintegration. Life and politics at the end of her rule will be unrecognizable from their grandeur and innocence at its beginning. "We don't blame her for it," Philip Ziegler, the historian and royal biographer, told me. "We have declined with her, so to speak."

The obituary films will remind us what a different country she inherited. One piece of footage will be played again and again: from her 21st birthday, in 1947, when Princess Elizabeth was on holiday with

"It is not unusual for a country to succumb to a state of denial as a long chapter in its history is about to end."

her parents in Cape Town. She was 6,000 miles from home and comfortably within the pale of the British Empire. The princess sits at a table with a microphone. The shadow of a tree plays on her shoulder. The camera adjusts three or four times as she talks, and on each occasion, she twitches momentarily, betraying tiny flashes of aristocratic irritation. "I declare before you all that my whole life, whether it be long or short, shall be devoted to your service, and the service of our great imperial family to which we all belong," she says, enunciating vowels and a conception of the world that have both vanished.

It is not unusual for a country to succumb to a state of denial as a long chapter in its history is about to end. When it became public that Queen Victoria was dying, at the age of 82, a widow for half her life, "astonished grief ... swept the country", wrote her biographer, Lytton Strachey. In the minds of her subjects, the queen's mortality had become unimaginable; and with her demise, everything was suddenly at risk, placed in the hands of an elderly and untrusted heir, Edward VII. "The wild waters are upon us now," wrote the American Henry James, who had moved to London 30 years before.

The parallels with the unease that we will feel at the death of Elizabeth II are obvious, but without the consolation of Britain's status in 1901 as the world's most successful country. "We have to have narratives for royal events," the historian told me. "In the Victorian reign, everything got better and better, and bigger and bigger. We certainly can't tell that story today."

The result is an enormous objection to even thinking about – let alone talking or writing about – what will happen when the Queen dies. We avoid the subject as we avoid it in our own families. It seems like good manners, but it is also fear. The reporting for this article involved dozens of interviews with broadcasters, government officials, and departed palace staff, several of whom have worked on London Bridge directly. Almost all insisted on complete secrecy. "This meeting never happened," I was told after one conversation in a gentleman's club on Pall Mall. Buckingham Palace, meanwhile, has a policy of not commenting on funeral arrangements for members of the royal family.

And yet this taboo, like much to do with the monarchy, is not entirely rational, and masks a parallel reality. The next great rupture in Britain's national life has, in fact, been planned to the minute. It involves matters of major public importance, will be paid for by us, and is definitely going to happen. According to the Office of National Statistics, a British woman who reaches the age of 91 – as the Queen will in April – has an average life expectancy of four years and three months. The Queen is approaching the end of her reign at a time of maximum disquiet about Britain's place in the world, at a moment when internal political tensions are close to breaking her kingdom apart. Her death will also release its own destabilising forces: in the accession of Queen Camilla; in the optics of a new king who is already an old man; and in the future of the Commonwealth, an invention largely of her making. (The Queen's title of "Head of the Commonwealth" is not hereditary.) Australia's prime minister and leader of the opposition both want the country to become a republic.

Coping with the way these events fall is the next great challenge of the House of Windsor, the last European royal family to practise coronations and to persist – with the complicity of a willing public – in the magic of the whole enterprise. That is why the planning for the Queen's death and its ceremonial aftermath is so extensive. Succession is part of the job. It is an opportunity for order to be affirmed. Queen Victoria had written down the contents of her coffin by 1875. The Queen Mother's funeral was rehearsed for 22 years. Louis Mountbatten, the last Viceroy of India, prepared a winter and a summer menu for his funeral lunch. London Bridge is the Queen's exit plan. "It's history," as one of her courtiers said. It will be 10 days of sorrow and spectacle in which, rather like the dazzling mirror of the monarchy itself, we will revel in who we were and avoid the question of what we have become.

The idea is for nothing to be unforeseen. If the Queen dies abroad, a BAe 146 jet from the RAF's No 32 squadron, known as the Royal Flight, will take off from Northolt, at the western edge of London, with a coffin on board. The royal undertakers, Leverton & Sons, keep what they call a "first call coffin" ready in case of royal emergencies. Both George V and George VI were buried in oak grown on the

Sandringham estate in Norfolk. If the Queen dies there, her body will come to London by car after a day or two.

The most elaborate plans are for what happens if she passes away at Balmoral, where she spends three months of the year. This will trigger an initial wave of Scottish ritual. First, the Queen's body will lie at rest in her smallest palace, at Holyroodhouse, in Edinburgh, where she is traditionally guarded by the Royal Company of Archers, who wear eagle feathers in their bonnets. Then the coffin will be carried up the Royal Mile to St Giles's cathedral, for a service of reception, before being put on board the Royal Train at Waverley station for a sad progress down the east coast mainline. Crowds are expected at level crossings and on station platforms the length of the country – from Musselburgh and Thirsk in the north, to Peterborough and Hatfield in the south – to throw flowers on the passing train. (Another locomotive will follow behind, to clear debris from the tracks.) "It's actually very complicated," one transport official told me.

In every scenario, the Queen's body returns to the throne room in Buckingham Palace, which overlooks the north-west corner of the Quadrangle, its interior courtyard. There will be an altar, the pall, the royal standard, and four Grenadier Guards, their bearskin hats inclined, their rifles pointing to the floor, standing watch. In the corridors, staff employed by the Queen for more than 50 years will pass, following procedures they know by heart. "Your professionalism takes over because there is a job to be done," said one veteran of royal funerals. There will be no time for sadness, or to worry about what happens next. Charles will bring in many of his own staff when he accedes. "Bear in mind," the courtier said, "everybody who works in the palace is actually on borrowed time."

Outside, news crews will assemble on pre-agreed sites next to Canada Gate, at the bottom of Green Park. (Special fibre-optic cable runs under the Mall, for broadcasting British state occasions.) "I have got in front of me an instruction book a couple of inches thick," said one TV director, who will cover the ceremonies, when we spoke on the phone. "Everything in there is planned. Everyone knows what to do." Across the country, flags will come down and bells will toll. In 1952, Great Tom was rung at St Paul's every minute for two hours when the news was announced. The bells at Westminster Abbey sounded and the Sebastopol bell, taken from the Black Sea city during the Crimean war and rung only on the occasion of a sovereign's death, was tolled 56 times at Windsor – once for each year of George VI's life – from 1.27pm until 2.22pm.

The 18th Duke of Norfolk, the Earl Marshal, will be in charge. Norfolks have overseen royal funerals since 1672. During the 20th century, a set of offices in St James's Palace was always earmarked for their use. On the morning of George VI's death, in 1952, these were being renovated. By five o'clock in the afternoon, the scaffolding was down and the rooms were re-carpeted, furnished and equipped with phones, lights and heating. During London Bridge, the Lord Chamberlain's office in the palace will be the centre of operations. The current version of the plan is largely the work of Lieutenant-Colonel Anthony Mather, a former equerry who retired from the palace in 2014. As a 23-year-old guardsman in 1965, Mather led the pallbearers at Churchill's funeral. (He declined to speak with me.) The government's team – coordinating the police, security, transport and armed forces – will assemble at the Department of Culture, Media and Sport. Someone will have the job of printing around 10,000 tickets for invited guests, the first of which will be required for the proclamation of King Charles in about 24 hours time.

Everyone on the conference calls and around the table will know each other. For a narrow stratum of the British aristocracy and civil service, the art of planning major funerals – the solemnity, the excessive detail – is an expression of a certain national competence. Thirty-one people gathered for the first meeting to plan Churchill's funeral, "Operation Hope Not", in June 1959, six years before his death. Those working on London Bridge (and Tay Bridge and Forth Bridge, the Duke of Edinburgh's funeral) will have corresponded for years in a language of bureaucratic euphemism, about "a possible future ceremony"; "a future problem"; "some inevitable occasion, the timing of which, however, is quite uncertain".

The first plans for London Bridge date back to the 1960s, before being refined in detail at the turn of the century. Since then, there have been meetings two or three times a year for the various actors involved (around a dozen government departments, the police, army, broadcasters and the Royal Parks) in Church House, Westminster, the Palace, or elsewhere in Whitehall. Participants described them to me as deeply civil and methodical. "Everyone around the world is looking to us to do this again perfectly," said one, "and we will." Plans are updated and old versions are destroyed. Arcane and highly specific knowledge is shared. It takes 28 minutes at a slow march from the doors of St James's to the entrance of Westminster Hall. The coffin must have a false lid, to hold the crown jewels, with a rim at least three inches high.

In theory, everything is settled. But in the hours after the Queen has gone, there will be details that only Charles can decide. "Everything has to be signed off by the Duke of Norfolk and the King," one official told me. The Prince of Wales has waited longer to assume the British throne than any heir, and the world will now swirl around him at a new and uncrossable distance. "For a little while," wrote Edward VIII, of the days between his father's death and funeral, "I had the uneasy sensation of being left alone on a vast stage." In recent years, much of the work on London Bridge has focused on the precise choreography of Charles's accession. "There are really two things happening," as one of his advisers told me. "There is the demise of a sovereign and then there is the making of a king." Charles is scheduled to make his first address as head of state on the evening of his mother's death.

Switchboards – the Palace, Downing Street, the Department of Culture, Media and Sport – will be swamped with calls during the first 48 hours. It is such a long time since the death of a monarch that many national organizations won't know what to do. The official advice, as it was last time, will be that business should continue as usual. This won't necessarily happen. If the Queen dies during Royal Ascot, the meet will be scrapped. The Marylebone Cricket Club is said to hold insurance for a similar outcome if she passes away during a home test match at Lord's. After the death of George VI in 1952, rugby and hockey fixtures were called off, while football matches went ahead. Fans sang Abide With Me and the national anthem before kickoff. The National Theatre will close if the news breaks before 4pm, and stay open if not. All games, including golf, will be banned in the Royal Parks.

In 2014, the National Association of Civic Officers circulated protocols for local authorities to follow in case of "the death of a senior national figure". It advised stockpiling books of condolence – loose leaf, so inappropriate messages can be removed – to be placed in town halls, libraries and museums the day after the Queen dies. Mayors will mask their decorations (maces will be shrouded with black bags). In provincial cities, big screens will be erected so crowds can follow events taking place in London, and flags of all possible descriptions, including beach flags (but not red danger flags), will be flown at half mast. The country must be seen to know what it is doing. The most recent set of instructions to embassies in London went out just before Christmas. One of the biggest headaches will be for the Foreign Office, dealing with all the dignitaries who descend from all corners of the earth. In Papua New Guinea, where the Queen is the head of state, she is known as "Mama belong big family". European royal families will be put up at the palace; the rest will stay at Claridge's hotel.

Parliament will gather. If possible, both houses will sit within hours of the monarch's death. In 1952, the Commons convened for two minutes before noon. "We cannot at this moment do more than record a spontaneous expression of our grief," said Churchill, who was prime minister. The house met again in the evening, when MPs began swearing the oath of allegiance to the new sovereign. Messages rained in from parliaments and presidents. The US House of Representatives adjourned. Ethiopia announced two weeks of mourning. In the House of Lords, the two thrones will be replaced by a single chair and a cushion bearing the golden outline of a crown.

On D+1, the day after the Queen's death, the flags will go back up, and at 11am, Charles will be proclaimed king. The Accession Council, which convenes in the red-carpeted Entrée Room of St James's Palace, long predates parliament. The meeting, of the "Lords Spiritual and Temporal of this Realm", derives from the Witan, the Anglo-Saxon feudal assembly of more than a thousand years ago. In theory, all 670 current members of the Privy Council, from Jeremy Corbyn to Ezekiel Alebua, the former prime minister of the Solomon Islands, are invited

– but there is space for only 150 or so. In 1952, the Queen was one of two women present at her proclamation.

The clerk, a senior civil servant named Richard Tilbrook, will read out the formal wording, "Whereas it has pleased Almighty God to call to His Mercy our late Sovereign Lady Queen Elizabeth the Second of Blessed and Glorious memory..." and Charles will carry out the first official duties of his reign, swearing to protect the Church in Scotland, and speaking of the heavy burden that is now his.

At dawn, the central window overlooking Friary Court, on the palace's eastern front, will have been removed and the roof outside covered in red felt. After Charles has spoken, trumpeters from the Life Guards, wearing red plumes on their helmets, will step outside, give three blasts and the Garter King of Arms, a genealogist named Thomas Woodcock, will stand on the balcony and begin the ritual proclamations of King Charles III. "I will make the first one," said Woodcock, whose official salary of £49.07 has not been raised since the 1830s. In 1952, four newsreel cameras recorded the moment. This time there will be an audience of billions. People will look for auguries – in the weather, in birds flying overhead – for Charles's reign. At Elizabeth's accession, everyone was convinced that the new queen was too calm. The band of the Coldstream Guards will play the national anthem on drums that are wrapped in black cloth.

The proclamations will only just be getting started. From St James's, the Garter King of Arms and half a dozen other heralds, looking like extras from an expensive Shakespeare production, will go by carriage to the statue of Charles I, at the base of Trafalgar Square, which marks London's official midpoint, and read out the news again. A 41-gun salute – almost seven minutes of artillery – will be fired from Hyde Park. "There is no concession to modernity in this," one former palace official told me. There will be cocked hats and horses everywhere. One of the concerns of the broadcasters is what the crowds will look like as they seek to record these moments of history. "The whole world is going to be bloody doing this," said one news executive, holding up his phone in front of his face.

On the old boundary of the City of London, outside the Royal Courts of Justice, a red cord will hang across the road. The City Marshal, a former police detective chief superintendent named Philip Jordan, will be waiting on a horse. The heralds will be formally admitted to the City, and there will be more trumpets and more announcements: at the Royal Exchange, and then in a chain reaction across the country. Sixty-five years ago, there were crowds of 10,000 in Birmingham; 5,000 in Manchester; 15,000 in Edinburgh. High Sheriffs stood on the steps of town halls, and announced the new sovereign according to local custom. In York, the Mayor raised a toast to the Queen from a cup made of solid gold.

The same rituals will take place, but this time around the new king will also go out to meet his people. From his proclamation at St James's, Charles will immediately tour the country, visiting Edinburgh, Belfast and Cardiff to attend services of remembrance for his mother and to meet the leaders of the devolved governments. There will also be civic receptions, for teachers, doctors and other ordinary folk, which are intended to reflect the altered spirit of his reign. "From day one, it is about the people rather than just the leaders being part of this new monarchy," said one of his advisers, who described the plans for Charles's progress as: "Lots of not being in a car, but actually walking around." In the capital, the pageantry of royal death and accession will be archaic and bewildering. But from another city each day, there will be images of the new king mourning alongside his subjects, assuming his almighty, lonely role in the public imagination. "It is see and be seen," the adviser said.

For a long time, the art of royal spectacle was for other, weaker peoples: Italians, Russians, and Habsburgs. British ritual occasions were a mess. At the funeral of Princess Charlotte, in 1817, the undertakers were drunk. Ten years later, St George's Chapel was so cold during the burial of the Duke of York that George Canning, the foreign secretary, contracted rheumatic fever and the bishop of London died. "We never saw so motley, so rude, so ill-managed a body of persons," reported the Times on the funeral of George IV, in 1830. Victoria's coronation a few years later was nothing to write home about. The clergy got lost in the words; the singing was awful; and the royal jewellers made the coronation ring for the wrong finger. "Some nations have a gift for ceremonial," the Marquess of Salisbury wrote in 1860. "In England the case is exactly the reverse."

What we think of as the ancient rituals of the monarchy were mainly crafted in the late 19th century, towards the end of Victoria's reign. Courtiers, politicians and constitutional theorists such as Walter Bagehot worried about the dismal sight of the Empress of India trooping around Windsor in her donkey cart. If the crown was going to give up its executive authority, it would have to inspire loyalty and awe by other means – and theatre was part of the answer. "The more democratic we get," wrote Bagehot in 1867, "the more we shall get to like state and show."

Obsessed by death, Victoria planned her own funeral with some style. But it was her son, Edward VII, who is largely responsible for reviving royal display. One courtier praised his "curious power of visualising a pageant". He turned the state opening of parliament and military drills, like the Trooping of the Colour, into full fancy-dress occasions, and at his own passing, resurrected the medieval ritual of lying in state. Hundreds of thousands of subjects filed past his coffin in Westminster Hall in 1910, granting a new sense of intimacy to the body of the sovereign. By 1932, George V was a national father figure, giving the first royal Christmas speech to the nation – a tradition that persists today – in a radio address written for him by Rudyard Kipling.

The shambles and the remoteness of the 19th-century monarchy were replaced by an idealised family and historic pageantry invented in the 20th. In 1909, Kaiser Wilhelm II boasted about the quality of German martial processions: "The English cannot come up to us in this sort of thing." Now we all know that no one else quite does it like the British.

The Queen, by all accounts a practical and unsentimental person, understands the theatrical power of the crown. "I have to be seen to be believed," is said to be one of her catchphrases. And there is no reason to doubt that her funeral rites will evoke a rush of collective feeling. "I think there will be a huge and very genuine outpouring of deep emotion," said Andrew Roberts, the historian. It will be all about her, and it will really be about us. There will be an urge to stand in the street, to see it with your own eyes, to be part of a multitude. The cumulative effect will be conservative. "I suspect the Queen's death will intensify patriotic feelings," one constitutional thinker told me, "and therefore fit the Brexit mood, if you like, and intensify the feeling that there is nothing to learn from foreigners."

The wave of feeling will help to swamp the awkward facts of the succession. The rehabilitation of Camilla as the Duchess of Cornwall has been a quiet success for the monarchy, but her accession as queen will test how far that has come. Since she married Charles in 2005, Camilla has been officially known as Princess Consort, a formulation that has no historical or legal meaning. ("It's bullshit," one former courtier told me, describing it as "a sop to Diana".) The fiction will end when Elizabeth II dies. Under common law, Camilla will become queen — the title always given to the wives of kings. There is no alternative. "She is queen whatever she is called," as one scholar put it. "If she is called Princess Consort there is an implication that she is not quite up to it. It's a problem." There are plans to clarify this situation before the Queen dies, but King Charles is currently expected to introduce Queen Camilla at his Accession Council on D+1. (Camilla was invited to join the Privy Council last June, so she will be present.) Confirmation of her title will form part of the first tumultuous 24 hours.

The Commonwealth is the other knot. In 1952, at the last accession, there were only eight members of the new entity taking shape in the outline of the British Empire. The Queen was the head of state in seven of them, and she was proclaimed Head of the Commonwealth to accommodate India's lone status as a republic. Sixty-five years later, there are 36 republics in the organization, which the Queen has attended assiduously throughout her reign, and now comprises a third of the world's population. The problem is that the role is not hereditary, and there is no procedure for choosing the next one. "It's a complete grey area," said Philip Murphy, director of the Institute of Commonwealth Studies at the University of London.

For several years, the palace has been discreetly trying to ensure Charles's succession as head of the bloc, in the absence of any other obvious option. Last October, Julia Gillard, the former prime minister of Australia, revealed that Christopher Geidt, the Queen's private secretary, had visited her in February 2013 to ask her to support the idea. Canada and New Zealand have since fallen into line, but the title is unlikely to be included in King Charles's proclamation. Instead it will be part of the discreet international lobbying that takes place as London fills up with diplomats and presidents in the days after the Queen's death. There will be serious, busy receptions at the palace. "We are not talking about entertaining. But you have to show some form of respect for the fact that they have come," said one courtier. "Such feasting and commingling, with my father still unburied, seemed to me unfitting and heartless," wrote Edward VIII in his memoirs. The show must go on. Business will mix with grief.

There will be a thousand final preparations in the nine days before the funeral. Soldiers will walk the processional routes. Prayers will be rehearsed. On D+1, Westminster Hall will be locked, cleaned and its stone floor covered with 1,500 meters of carpet. Candles, their wicks already burnt in, will be brought over from the Abbey. The streets around will be converted into ceremonial spaces. The bollards on the Mall will be removed, and rails put up to protect the hedges. There is space for 7,000 seats on Horse Guards Parade and 1,345 on Carlton House Terrace. In 1952, all the rhododendrons in Parliament Square were pulled up and women were barred from the roof of Admiralty Arch. "Nothing can be done to protect the bulbs," noted the Ministry of Works. The Queen's 10 pallbearers will be chosen, and practice carrying their burden out of sight in a barracks somewhere. British royals are buried in lead-lined coffins. Diana's weighed a quarter of a ton.

The population will slide between sadness and irritability. In 2002, 130 people complained to the BBC about its insensitive coverage of the Queen Mother's death; another 1,500 complained that Casualty was moved to BBC2. The TV schedules in the days after the Queen's death will change again. Comedy won't be taken off the BBC completely, but most satire will. There will be Dad's Army reruns, but no Have I Got News For You.

People will be touchy either way. After the death of George VI, in a society much more Christian and deferential than this one, a Mass Observation survey showed that people objected to the endless maudlin music, the forelock-tugging coverage. "Don't they think of old folk, sick people, invalids?" one 60-year old woman asked. "It's been terrible for them, all this gloom." In a bar in Notting Hill, one drinker said, "He's only shit and soil now like anyone else," which started a fight. Social media will be a tinderbox. In 1972, the writer Brian Masters estimated that around a third of us have dreamed about the Queen – she stands for authority and our mothers. People who are not expecting to cry will cry.

On D+4, the coffin will move to Westminster Hall, to lie in state for four full days. The procession from Buckingham Palace will be the first great military parade of London Bridge: down the Mall, through Horse Guards, and past the Cenotaph. More or less the same slow march, from St James's Palace for the Queen Mother in 2002, involved 1,600 personnel and stretched for half a mile. The bands played Beethoven and a gun was fired every minute from Hyde Park. The route is thought to hold around a million people. The plan to get them there is based on the logistics for the London 2012 Olympics.

There may be corgis. In 1910, the mourners for Edward VII were led by his fox terrier, Caesar. His son's coffin was followed to Wolferton station, at Sandringham, by Jock, a white shooting pony. The procession will reach Westminster Hall on the hour. The timing will be just so. "Big Ben beginning to chime as the wheels come to a stop," as one broadcaster put it.

Inside the hall, there will be psalms as the coffin is placed on a catafalque draped in purple. King Charles will be back from his tour of the home nations, to lead the mourners. The orb, the sceptre and the Imperial Crown will be fixed in place, soldiers will stand guard and then the doors opened to the multitude that will have formed outside and will now stream past the Queen for 23 hours a day. For George VI, 305,000 subjects came. The line was four miles long. The palace is expecting half a million for the Queen. There will be a wondrous queue – the ultimate British ritual undertaking, with canteens, police, portable toilets and strangers talking cautiously to one another – stretching down to Vauxhall Bridge and then over the river and back along the Albert Embankment. MPs will skip to the front.

Under the chestnut roof of the hall, everything will feel fantastically well-ordered and consoling and designed to within a quarter of an inch, because it is. A 47-page internal report compiled after George VI's funeral suggested attaching metal rollers to the catafalque, to smooth the landing of the coffin when it arrives. Four soldiers will stand silent vigil for 20 minutes at a time, with two ready in reserve. The RAF, the Army, the Royal Navy, the Beefeaters, the Gurkhas – everyone will take part. The most senior officer of the four will stand at the foot of the coffin, the most junior at the head. The wreaths on the coffin will be renewed every day. For Churchill's lying in state in 1965, a replica of the hall was set up in the ballroom of the St Ermin's hotel nearby, so soldiers could practise their movements before they went on duty. In 1936, the four sons of George V revived The Prince's Vigil, in which members of the royal family arrive unannounced and stand watch. The Queen's children and grandchildren – including women for the first time – will do the same.

Before dawn on D+9, the day of the funeral, in the silent hall, the jewels will be taken off the coffin and cleaned. In 1952, it took three jewellers almost two hours to remove all the dust. (The Star of Africa, on the royal sceptre, is the second-largest cut diamond in the world.) Most of the country will be waking to a day off. Shops will close, or go to bank holiday hours. Some will display pictures of the Queen in their windows. The stock market will not open. The night before, there will have been church services in towns across the UK. There are plans to open football stadiums for memorial services if necessary.

At 9am, Big Ben will strike. The bell's hammer will then be covered with a leather pad seven-sixteenths of an inch thick, and it will ring out in muffled tones. The distance from Westminster Hall to the Abbey is only a few hundred meters. The occasion will feel familiar, even though it is new: the Queen will be the first British monarch to have her funeral in the Abbey since 1760. The 2,000 guests will be sitting inside. Television cameras, in hides made of painted bricks, will search for the images that we will remember. In 1965, the dockers dipped

"At 9am, Big Ben will strike. The bell's hammer will then be covered with a leather pad seven-sixteenths of an inch thick, and it will ring out in muffled tones."

their cranes for Churchill. In 1997, it was the word "Mummy" on the flowers for Diana from her sons.

When the coffin reaches the abbey doors, at 11 o'clock, the country will fall silent. The clatter will still. Train stations will cease announcements. Buses will stop and drivers will get out at the side of the road. In 1952, at the same moment, all of the passengers on a flight from London to New York rose from their seats and stood, 18,000 feet above Canada, and bowed their heads.

Back then, the stakes were clearer, or at least they seemed that way. A stammering king had been part of the embattled British way of life that had survived an existential war. The wreath that Churchill laid said: "For Gallantry." The BBC commentator in 1952, the man who deciphered the rubies and the rituals for the nation, was Richard Dimbleby, the first British reporter to enter Bergen-Belsen and convey its horrors, seven years before. "How true tonight that statement spoken by an unknown man of his beloved father," murmured Dimbleby, describing the lying in state to millions. "The sunset of his death tinged the whole world's sky."

The trumpets and the ancientness were proof of our survival; and the king's young daughter would rule the peace. "These royal ceremonies represented decency, tradition, and public duty, in contradiction to the ghastliness of Nazism," as one historian told me. The monarchy had traded power for theatre, and in the aftermath of war, the illusion became more powerful than anyone could have imagined. "It was restorative," Jonathan Dimbleby, Richard's son and biographer, told me.

His brother, David, is likely to be behind the BBC microphone this time. The question will be what the bells and the emblems and the heralds represent now. At what point does the pomp of an imperial monarchy become ridiculous amid the circumstances of a diminished nation? "The worry," a historian said, "is that it is just circus animals."

If the monarchy exists as theatre, then this doubt is the part of the drama. Can they still pull it off? Knowing everything that we know in 2017, how can it possibly hold that a single person might contain the soul of a nation? The point of the monarchy is not to answer such questions. It is to continue. "What a lot of our life we spend in acting," the Queen Mother used to say.

Inside the Abbey, the archbishop will speak. During prayers, the broadcasters will refrain from showing royal faces. When the coffin emerges again, the pallbearers will place it on the green gun carriage that was used for the Queen's father, and his father and his father's father, and 138 junior sailors will drop their heads to their chests and pull. The tradition of being hauled by the Royal Navy began in 1901 when Victoria's funeral horses, all white, threatened to bolt at Windsor Station and a waiting contingent of ratings stepped in to pull the coffin instead.

The procession will swing on to the Mall. In 1952, the RAF was grounded out of respect for King George VI. In 2002, at 12.45pm, a Lancaster bomber and two Spitfires flew over the cortege for his wife and dipped their wings. The crowds will be deep for the Queen. She will get everything. From Hyde Park Corner, the hearse will go 23 miles by road to Windsor Castle, which claims the bodies of British sovereigns. The royal household will be waiting for her, standing on the grass. Then the cloister gates will be closed and cameras will stop broadcasting. Inside the chapel, the lift to the royal vault will descend, and King Charles will drop a handful of red earth from a silver bowl.

This Is What a 21st-Century Police State Really Looks Like

By Megha Rajagopalan

(First appeared in
Buzzfeed News,
October 2017)

K ASHGAR, China — This is a city where growing a beard can get you reported to the police. So can inviting too many people to your wedding, or naming your child Muhammad or Medina.

Driving or taking a bus to a neighboring town, you'd hit checkpoints where armed police officers might search your phone for banned apps like Facebook or Twitter, and scroll through your text messages to see if you had used any religious language.

You would be particularly worried about making phone calls to friends and family abroad. Hours later, you might find police officers knocking at your door and asking questions that make you suspect they were listening in the whole time.

For millions of people in China's remote far west, this dystopian future is already here. China, which has already deployed the world's most sophisticated internet censorship system, is building a surveillance state in Xinjiang, a four-hour flight from Beijing, that uses both the newest technology and human policing to keep tabs on every aspect of citizens' daily lives. The region is home to a Muslim ethnic minority called the Uighurs, who China has blamed for forming separatist groups and fueling terrorism. Since this spring, thousands of Uighurs and other ethnic minorities have disappeared into so-called political education centers, apparently for offenses from using Western social media apps to studying abroad in Muslim countries, according to relatives of those detained.

Over the past two months, I interviewed more than two dozen Uighurs, including recent exiles and those who are still in Xinjiang, about what it's like to live there. The majority declined to be named because they were afraid that police would detain or arrest their families if their names appeared in the press.

Taken along with government and corporate records, their accounts paint a picture of a regime that at once recalls the paranoia of the Mao era and is also thoroughly modern, marrying heavy-handed human policing of any behavior outside the norm with high-tech tools like iris recognition and apps that eavesdrop on cell phones.

China's government says the security measures are necessary in Xinjiang because of the threat of extremist violence by Uighur militants — the region has seen periodic bouts of unrest, from riots in 2009 that left almost 200 dead to a series of deadly knife and bomb attacks in 2013 and 2014. The government also says it's made life for Uighurs better, pointing to the money it's poured into economic development in the region, as well as programs making it easier for Uighurs to attend university and obtain government jobs. Public security and propaganda authorities in Xinjiang did not respond to requests for comment. China's Foreign Ministry said it had no knowledge of surveillance measures put in place by the local government.

"I want to stress that people in Xinjiang enjoy a happy and peaceful working and living situation," said Lu Kang, a spokesperson for China's Foreign Ministry, when asked why the surveillance measures are needed. "We have never heard about these measures taken by local authorities."

But analysts and rights groups say the heavy-handed restrictions punish all of the region's 9 million Uighurs — who make up a bit under half of the region's total population — for the actions of a handful of people. The curbs themselves fuel resentment and breed extremism, they say.

The ubiquity of government surveillance in Xinjiang affects the most prosaic aspects of daily life, those interviewed for this story said. D., a stylish young Uighur woman in Turkey, said that even keeping in touch with her grandmother, who lives in a small Xinjiang village, had become impossible.

Whenever D. called her grandmother, police would barge in hours later, demanding the elderly woman phone D. back while they were in the room.

"For god's sake, I'm not going to talk to my 85-year-old grandmother about how to destroy China!" D. said, exasperated, sitting across the table from me in a café around the corner from her office.

After she got engaged, D. invited her extended family, who live in Xinjiang, to her wedding. Because it is now nearly impossible for Uighurs to obtain passports, D. ended up postponing the ceremony for months in hopes the situation would improve.

Finally, in May, she and her mother had a video call with her family on WeChat, the popular Chinese messaging platform. When D. asked how they were, they said everything was fine. Then one of her relatives, afraid of police eavesdropping, held up a handwritten sign that said, "We could not get the passports."

D. felt her heart sink, but she just nodded and kept talking. As soon as the call ended, she said, she burst into tears.

"Don't misunderstand me, I don't support suicide bombers or anyone who attacks innocent people," she said. "But in that moment, I told my mother I could understand them. I was so pissed off that I could understand how those people could feel that way."

China's government has invested billions of renminbi into top-of-the-line surveillance technology for Xinjiang, from facial recognition cameras at petrol stations to surveillance drones that patrol the border.

China is not alone in this — governments from the United States to Britain have poured funds into security technology and know-how to combat threats from terrorists. But in China, where Communist Party–controlled courts convict 99.9% of the accused and arbitrary detention is a common practice, digital and physical spying on Xinjiang's populace has resulted in disastrous consequences for Uighurs and other ethnic minorities. Many have been jailed after they advocated for more rights or extolled Uighur culture and history, including the prominent scholar Ilham Tohti.

China has gradually increased restrictions in Xinjiang for the past decade in response to unrest and violent attacks, but the surveillance has been drastically stepped up since the appointment of a new party boss to the region in August 2016. Chen Quanguo, the party secretary, brought "grid-style social management" to Xinjiang, placing police and paramilitary troops every few hundred feet and establishing thousands of "convenience police stations." The use of political education centers — where thousands have been detained this year without charge — also radically increased after his tenure began. Spending

"'People disappear inside that place,' said the owner of a business in the area. 'So many people — many of my friends.'"

on domestic security in Xinjiang rose 45% in the first half of this year, compared to the same period a year earlier, according to an analysis of Chinese budget figures by researcher Adrian Zenz of the European School of Culture and Theology in Germany. A portion of that money has been poured into dispatching tens of thousands of police officers to patrol the streets.

In an August speech, Meng Jianzhu, China's top domestic security official, called for the use of a DNA database and "big data" in keeping Xinjiang secure.

It's a corner of the country that has become a window into the possible dystopian future of surveillance technology, wielded by states like China that have both the capital and the political will to monitor — and repress — minority groups. The situation in Xinjiang could be a harbinger for draconian surveillance measures rolled out in the rest of the country, analysts say.

"It's an open prison," said Omer Kanat, director of the Washington-based Uyghur Human Rights Project, an advocacy group that conducts research on life for Uighurs in Xinjiang. "The Cultural Revolution has returned [to the region], and the government doesn't try to hide anything. It's all in the open."

Once an oasis town on the ancient Silk Road, Kashgar is the cultural heart of the Uighur community. On a sleepy tree-lined street in the northern part of the city, among noodle shops and bakeries, stands an imposing compound surrounded by high concrete walls topped with loops of barbed wire. The walls are papered with colorful posters bearing slogans like "cherish ethnic unity as you cherish your own eyes" and "love the party, love the country."

The compound is called the Kashgar Professional Skills Education and Training Center, according to a sign posted outside its gates. When I took a cell phone photo of the sign in September, a police officer ran out of the small station by the gate and demanded I delete it.

"What kind of things do they teach in there?" I asked.

"I'm not clear on that. Just delete your photo," he replied.

"People disappear inside that place."

Before this year, the compound was a school. But according to three people with friends and relatives held there, it is now a political education center — one of hundreds of new facilities where Uighurs are held, frequently for months at a time, to study the Chinese language, Chinese laws on Islam and political activity, and all the ways the Chinese government is good to its people.

"People disappear inside that place," said the owner of a business in the area. "So many people — many of my friends."

He hadn't heard from them since, he said, and even their families cannot reach them. Since this spring, thousands of Uighurs and other minorities have been detained in compounds like this one. Though the centers aren't new, their purpose has been significantly expanded in Xinjiang over the last few months.

Through the gaps in the gates, I could see a yard decorated with a white statue in the Soviet-era socialist realist style, a red banner bearing a slogan, and another small police station. The beige building inside had shades over each of its windows.

Chinese state media has acknowledged the existence of the centers, and often boasts of the benefits they confer on the Uighur populace. In an interview with the state-owned Xinjiang Daily, a 34-year-old Uighur farmer, described as an "impressive student," says he never realized until receiving political education that his behavior and style of dress could be manifestations of "religious extremism."

Detention for political education of this kind is not considered a form of criminal punishment in China, so no formal charges or sentences are given to people sent there, or to their families. So it's hard to say exactly what transgressions prompt authorities to send people to the centers. Anecdotal reports suggest that having a relative who has been convicted of a crime, having the wrong content on your cell phone, and appearing too religious could all be causes.

It's clear, though, that having traveled abroad to a Muslim country, or having a relative who has traveled abroad, puts people at risk of detention. And the ubiquity of digital surveillance makes it nearly impossible to contact relatives abroad, according to the Uighurs I interviewed.

One recent exile reported that his wife, who remained in Xinjiang with their young daughter, asked for a divorce so that police would stop questioning her about his activities.

"It's too dangerous to call home," said another Uighur exile in the Turkish capital, Ankara. "I used to call my classmates and relatives. But then the police visited them, and the next time, they said, 'Please don't call anymore.'"

R., a Uighur student just out of undergrad, discovered he had a knack for Russian language in college. He was dying to study abroad. Because of the new rules imposed last year that made it nearly impossible for Uighurs to obtain passports, the family scraped together about 10,000 RMB ($1500) to bribe an official and get one, R. said.

R. made it to a city in Turkey, where he started learning Turkish and immersed himself in the culture, which has many similarities to Uighur customs and traditions. But he missed his family and the cotton farm they run in southern Xinjiang. Still, he tried to avoid calling home too much so he wouldn't cause them trouble.

"She would never talk like that. It felt like a police officer was standing next to her."

"In the countryside, if you get even one call from abroad, they will know. It's obvious," said R., who agreed to meet me in the back of a trusted restaurant only after all the other patrons had gone home for the night. He was so nervous as he spoke that he couldn't touch the lamb-stuffed pastries on his plate.

In March, R. told me, he found out that his mother had disappeared into a political education center. His father was running the farm alone, and no one in the family could reach her. R. felt desperate.

Two months later, he finally heard from his mother. In a clipped phone call, she told him how grateful she was to the Chinese Communist Party, and how good she felt about the government.

"I know she didn't want to say it. She would never talk like that," R. said. "It felt like a police officer was standing next to her."

Since that call, his parents' phones have been turned off. He hasn't heard from them since May.

Security has become a big business opportunity for hundreds of companies, mostly Chinese, seeking to profit from the demand for surveillance equipment in Xinjiang.

Researchers have found that China is pouring money into its budget for surveillance. Zenz, who has closely watched Xinjiang's government spending on security personnel and systems, said its investment in information technology transfer, computer services, and software will quintuple this year from 2013. The growth in the security industry there reflects the state-backed surveillance boom, he said.

He noted that a budget line item for creating a "shared information platform" appeared for the first time this year. The government has also hired tens of thousands more security personnel.

Armed police, paramilitary forces, and volunteer brigades stand on every street in Kashgar, stopping pedestrians at random to check their identifications, and sometimes their cell phones, for banned apps like WhatsApp as well as VPNs and messages with religious or political content.

Other equipment, like high-resolution cameras and facial recognition technology, is ubiquitous. In some parts of the region, Uighurs have been made to download an app to their phones that monitors their messages. Called Jingwang, or "web cleansing," the app works to monitor "illegal religious" content and "harmful information," according to news reports.

The internet is painfully slow in the region. Maya Wang, a China researcher for Human Rights Watch, has also documented the use of a DNA database targeting Uighurs as well as political dissidents and migrants, along with the use of voice pattern recognition.

When I walked into a checkpoint a few miles east of Kashgar, a police officer stood near the entrance to check commuters' cell phones for banned apps and messages (as a foreigner I was sent to a separate line and not asked for my phone). Their faces were then scanned by a facial recognition camera and matched with their identification cards. Glossy white machines for full-body scans stood on the other side of the room.

Petrol stations have a similar setup. At a station I visited in Kashgar in September, visitors were stepping out of their cars to have their faces scanned and matched with identity cards before filling up. As a foreigner, I was only asked for my passport.

"Because Xinjiang is not stable, we have been able to sell a lot to the government authorities there," said the owner of a small Beijing-based company that manufactures surveillance drones and other security equipment for Chinese law enforcement agencies. Xinjiang, he said, uses the drones near its western borders.

The government relies on contracting with companies like Beijing Wanlihong Technology Company, which produces an iris-recognition system that it says is more accurate than facial and fingerprint scanning techniques. Wanlihong is involved in a pilot project in Kashgar that includes providing equipment and training.

"The goal of the system is to build a powerful and extensive identity verification system to identify key suspects and initiate an emergency response mechanism in a timely way," the company says on its website.

The data could be collected and used to monitor the physical movements of suspicious people on roads, it adds, or combined with their SMS and browsing data obtained from cellular carriers.

Urumqi-based Leon Technology — a company that integrates artificial intelligence into its security services and then provides those services to telecommunications companies and government agencies in Xinjiang and elsewhere in China — saw its earnings grow by 260% in the first quarter of 2017.

"The Xinjiang government is bound to spend large sums of money to safeguard people's property and the safety of their lives, protecting the region's peace, development and stability," it said in an article posted on its website.

James Leibold, an associate professor at La Trobe University in Australia who is conducting research on security contractors in Xinjiang, said the broader security industry, including both physical policing and surveillance, is now the biggest employer of people in the region.

"Marketers in Shanghai are calling it the golden era of investment in security in Xinjiang," he said.

Leibold noted that despite the high-tech equipment, it's still unclear how effective the government is at actually analyzing the large volumes of video and audio recordings. But it's clear the state is now pouring resources into doing just that — marrying information gleaned from phone tapping, security cameras, and other sources in an effort to create detailed profiles of people.

"In some ways it's like a high-tech version of the Cultural Revolution, like the social intrusion aspect and the regulations on religious behavior," Zenz said, referring to the Mao-era movement that's remembered in part for pitting neighbors against each other in a violent campaign to punish those seen as enemies of the party. "But the comparison breaks down because it's systematic and deliberate, and low-key — the Cultural Revolution caused a big mess."

"This is entirely top-down control," he added.

"It's a kind of frontline laboratory for surveillance."

State-owned companies are using Xinjiang as a testing ground for big data, Zenz said, and Xinjiang has historically been used to test out surveillance technology that is later rolled out in other parts of the country. Many companies have set up R&D labs in the region for this purpose with government backing.

"It's a kind of frontline laboratory for surveillance," Zenz said. "Because it's a bit outside of the public eye, there can be more experimentation there."

Surveillance in Xinjiang may be particularly harsh, but it's clear the government is expanding the use of the technology in the rest of the country, too. Outside Xinjiang, facial recognition is already being used increasingly by Chinese law enforcement to catch criminals, according to media reports. At a beer festival in the seaside city of Qingdao in August, 49 people found themselves arrested when cameras matched their faces with a national police database that showed they were suspected of crimes like theft and drug use.

"Those wanted criminals let their guard down when they went to the festival, which doesn't check for ID," a local police official told online news outlet Sixth Tone. "But they were not aware that a simple shot of their faces would lead to their arrest."

In another big data foray, the government is also working on a "social credit system," which would put together lots of variables, including patriotism and moral behavior, to assign numerical scores to its citizens.

But beyond digital surveillance, many said the government has simply flooded the region with personnel dedicated to tracking residents' every move.

T., a writer, lived in an apartment complex in the regional capital of Urumqi with his wife and daughters until the middle of this summer. (He and his family are now in the US. He asked me not to disclose which city because he was afraid of being identified by the government.) For years, an official representing the neighborhood's Communist Party committee would visit his home every week and ask a set of questions that soon became mundane: Who had come to visit? Was anyone pregnant? Had anyone changed jobs? She would then report the information to the local police department, he said.

Then in April, the questions changed. The official began to ask whether the family was Muslim, and how they practiced. T. had never been very religious. But he says he respected Islam because it's a big part of Uighur culture. The family kept a small collection of religious texts on their bookshelves, as well as four prayer rugs. But the questions made him nervous. He told the official he was not a believer.

A month later, the disappearances started. Friends would vanish in the middle of the night, spirited away by police to political education centers. His neighbors began to disappear, he said, one after the other. T. was terrified.

Every evening he placed an overcoat and a pair of thick winter trousers near the door so he could pull them on quickly if the police came for him — the weather was warm but he was afraid he could be held into the winter months. He gave away the prayer rugs, and in the relative safety of the apartment, he burned every religious book.

"My wife was so upset, she told me, 'You can't do that,'" he said. "I told her, what choice do I have? If someone saw them in a public trash bin, it could bring us so much trouble."

The first people in T.'s apartment building to disappear, he said, were those who had traveled abroad and returned, particularly to Muslim countries, from Malaysia to Egypt. Then, in June, he says the police began to conduct random checks of pedestrians' mobile phones at street corners, bus stops, and petrol stations, sometimes downloading their contents to handheld devices.

The police would dispense warnings to anyone whose phone carried banned apps like WhatsApp and Facebook. Sometimes, he said, police would come to some people's homes and businesses to check their computers for banned software and content.

"If they find anything in there, it'll be trouble for you," he said. "It was a new kind of police — the internet police."

Checkpoints have made Abduweli Ayup nervous ever since he was released from prison. The ordeal had been devastating for the bookish 43-year-old, who was jailed in 2013 after he worked to set up kindergartens and other schools teaching children in the Uighur language. (He was formally charged with illegal fundraising.)

After 15 months in prison, Abduweli returned home to his wife and two young daughters, and he was hoping things could go back to normal. But one July day in 2015, trouble found him again as he was commuting to work at a checkpoint he must have been through a thousand times.

The officers who usually waved him through had been replaced by Special Weapons and Tactics police. When they saw his ID papers, they noticed his prison history.

That's when they demanded his laptop.

"I said, 'They know me here. I come here every day,'" Abduweli said. In response, one of the officers slapped him across the face, he said. When the police opened his laptop, they found essays he had written, years before, during an academic fellowship in Kansas. In his writing, he expressed his views on taboo subjects from Uighur culture to dictatorships as a system of government.

Abduweli was detained immediately, strip searched, and interrogated for hours about his writing by a group of six officers, he said.

One of the officers told him if he was caught with essays like that on his laptop again, he would be sent back to prison.

"That's when I decided the law here does not exist," Abduweli said, recounting the story in September at a friend's apartment in Ankara, where he fled later that year. "I realized if they took my computer again, it would be dangerous. So I knew I had to leave."

The government has said repeatedly that its goal in Xinjiang is to achieve both ethnic unity and social stability. Beyond curbs on the Uighur population, it has also provided subsidies for some Uighur workers and affirmative action programs for students to attend top universities, and takes pains to recruit them into government jobs. A significant percentage of the police in Kashgar, for instance, are themselves Uighur.

China also points to its growing investment in the infrastructure and key industries in historically underdeveloped Xinjiang, including roads, construction, and telecommunications networks.

But critics say those efforts are overshadowed by the government's more repressive measures, which have fueled the propaganda of a small number of extremists, including the Turkestan Islamic Party (TIP), a radical Uighur militant group, some of whose adherents are believed to have been active in Syria and Afghanistan. According to a translation by the Middle East Media Research Institute, TIP released an issue of its online magazine on Aug. 29, calling for its fighters in Syria to prepare for future battle against the Chinese state in East Turkestan, a name many Uighurs use to refer to Xinjiang. The issue sought to draw attention to the situation in Xinjiang and the treatment of Muslims by the Chinese government.

"I can see it in the younger generation," said D., the woman who postponed her wedding because her relatives in Xinjiang couldn't obtain passports in order to attend. "They are more angry."

These days, Abduweli is trying to finish a children's book explaining Uighur language and culture to elementary school kids. The pages are illustrated with clip art of generic cartoon children, culled from a Google image search. What the book really needs is a Uighur illustrator who understands the culture, Abduweli said, but because of his status as a former political prisoner, all the really good artists are too afraid to be publicly associated with him.

In December, the Chinese government abruptly canceled Abduweli's passport, rendering him stateless. He's applying for refugee status through the UN in Ankara.

Sometimes, he thinks back to his college years in Beijing in the early 1990s, which he remembers as his first real taste of freedom.

Abduweli remembers vividly a Chinese-language book he picked up from a street vendor. He loved it so much he biked across the city to a foreign-language bookshop to find the original English version, which he scoured page by page. He felt sure there were parts that the Chinese censors had removed.

The book was George Orwell's *1984,* and it reminded him, he said, of home.

How a Tax Haven Is Leading the Race to Privatize Space

By Atossa Araxia Abrahamian

(Originally appeared in
the *Guardian*,
September 2017)

On a drizzly afternoon in April, Prince Guillaume, the hereditary grand duke of Luxembourg, and his wife, Princess Stéphanie, sailed through the front doors of an office building in the outskirts of Seattle and into the headquarters of an asteroid-mining startup called Planetary Resources, which plans to "expand the economy into space".

The company's engineers greeted the royals with hors d'oeuvres, craft beer and bottles upon bottles of Columbia Valley rieslings and syrahs. In the corner of the lounge stood a vintage Asteroids arcade game; on the wall hung an American flag alongside the grand duchy's own red, white and blue stripes. Between the two flags was a prototype of a spacecraft designed to roam the galaxy, prospecting asteroids for precious natural resources that would someday – at least in theory – make the shareholders of Planetary Resources very wealthy earthlings indeed.

The nation of Luxembourg is one of Planetary Resources' main boosters. The country's pledge of €25m (£22.5m) – which includes both direct funding and state support for research and development – is just one element of its wildly ambitious campaign to become a terrestrial hub for the business of mining minerals, metals and other resources on celestial bodies. The tiny country enriched itself significantly over the past century by greasing the wheels of global finance; now, as companies such as Planetary Resources prepare for a cosmic land grab, Luxembourg wants to use its tiny terrestrial perch to help send capitalism into space.

Space exploration has historically been an arena for grand, nationalistic operations that were too costly, dangerous and complex for civilians to take up without state backing. But now, private companies want in, raising questions that, until recently, have seemed like mere thought experiments or hypotheticals: who can lay claim to an

"Who can lay claim to an asteroid and all of its extractive wealth?"

asteroid and all of its extractive wealth? Should space benefit "all of humankind", as the international treaties signed in the 60s intended, or is that idealism outdated? How do you measure those benefits, anyway? Does trickle-down theory apply in zero-gravity conditions?

Space is becoming a testing ground for these thorny ethical and legal questions, and Luxembourg – a tiny country that has sustained itself off of regulatory intricacies and tax loopholes for decades – is positioning itself to help find the answers. While major nations such as China and India plough increasing sums of money into developing space programmes to rival Nasa, Luxembourg is making a different bet: that it can become home to a multinational cast of entrepreneurs who want to go into space not for just the sake of scientific progress or to strengthen their nation's geopolitical hand, but also to make money.

It already has a keen clientele. Space entrepreneurs speak of a new "gold rush" and compare their mission to that of the frontiersmen, or the early industrialists. While planet Earth's limited stock of natural resources is rapidly being depleted, asteroid miners see a solution in the vast quantities of untapped water, minerals and metals in outer space. And the fledgling "NewSpace" industry – an umbrella term for commercial spaceflight, asteroid mining and other private ventures – has found eager supporters in the investor class. In April, Goldman Sachs sent a note to clients claiming that asteroid mining "could be more realistic than perceived", thanks to the falling cost of launching rockets and the vast quantities of platinum sitting on space rocks, just waiting to be exploited.

"[Mining asteroids] is not a new idea, but what's new is state support of the idea," says Chris Voorhees, the chief engineer of Planetary Resources. "Everyone thought it was inevitable but they weren't sure when it would occur." Now, he says, Luxembourg is "making it happen".

The grand duchy – which has all the square footage of an asteroid and, with a population of half a million, not all that many more inhabitants – has earmarked €200m to fund NewSpace companies that join its new space sector; to date, six have taken it up on the offer. It has sent officials to Japan, China and the UAE to talk about space exploration partnerships, and appointed space industry veterans, including the ex-head of the European Space Agency, to advise them. In May, it took out a glossy supplement in Scientific American magazine to signal it is committed not just to helping businesses, but to advancing research as well.

And in July, the parliament passed its law – the first of its kind in Europe, and the most far-reaching in the world – asserting that if a Luxembourgish company launches a spacecraft that obtains water, silver, gold or any other valuable substance on a celestial body, the extracted materials will be considered the company's legitimate private property by a legitimate sovereign nation.

The presence of royalty at Planetary Resources HQ ahead of the passing of the law was a canny part of the country's space incursion. The young couple was there to dazzle, charm and lend gravitas to the operation – European aristocracy doesn't show up in suburban office parks any old day – but the mission's greater aim was to impress upon Silicon Valley executives, the bemused Luxembourgish press and space scientists around the world that mining asteroids was no longer science fiction. To that end, the royals were accompanied by about 40 of their subjects, all of whom had a role to play in this emerging industry.

Etienne Schneider, Luxembourg's congenial deputy prime minister, led the delegation. With his easy manner, excellent English and penchant for fancy cars, he cuts a Macronian figure: a product of European socialist political parties, sure, and a social liberal to his core – Schneider is married to a man – but one who will willingly play handmaiden to global capitalist interests should the right opportunity arise. He announced recently that he would be running for the role of prime minister in 2018.

With Schneider came a delegation of scientists, trade attaches, bankers, lawyers and local journalists who switched between German, English, French and the local language, a consonant-heavy mix of Flemish and German with the occasional foreign word thrown in to supplement: "meeting", "framework", "brunch". ("We don't have all the words," a member of the delegation told me sheepishly.) In French, the language is known as Luxembourgeois, which pretty much says it all; the duchy's 500,000 citizens, who have a GDP per capita of $104,000 (£78,800), are the wealthiest in the world after Qatar's, according to the International Monetary Fund.

The Planetary Resources team took their benefactors on a tour of the labs where its hardware is built. The company isn't mining asteroids yet, but to benefit from Luxembourg's concessions, it opened an office in the grand duchy this year. Up close, its Arkyd 6 spacecraft – which is ready for launch – looks just like satellites look in the movies, only smaller. It had multiple flaps and appendages, including an infrared sensor, a star tracker to orient the craft in space and a GPS unit, which works only in the earth's orbit.

Once the tour was complete, cocktail hour began. Schneider, who owns a vineyard, bounced from one conversation to another, brimming with enthusiasm. To end the visit, Chris Lewicki, the CEO of the company, gave a toast praising Luxembourg's contributions "to an abundant future for all of humanity". As a parting gift, he presented her royal highness with a necklace. Instead of jewels, it was studded with tiny fragments of asteroids.

It is reasonable to wonder what, exactly, a marginal European monarchy, egged on by a vivacious gay socialist, was doing telling American entrepreneurs on the cutting edge of innovation that their hamlet-sized state could propel humanity – and capitalism – into deep space. The grand duchy has no national space agency, no launching sites, and only modest research capabilities. It opened its first and only university in 2003 and its military consists of 1,008 troops. Luxembourg does not fit the image of a spacefaring nation; in fact, some have questioned whether it should even be a nation at all.

Yet Luxembourg's very essence – as a speck in the heart of Europe – allows, even requires, it to partake in such ambitious ventures. Its national motto is "We want to remain what we are" and, over the centuries, this independent spirit has endured occupations by the dukes of Burgundy, the kings of Spain and France, the emperors of Austria and the king of the Netherlands. Today, the state, which only gained full independence in 1867, occupies a curious position in the global imagination: a country with an outsized economic influence that everyone has heard of, but that no one can quite locate on a map.

According to Gabriel Zucman, assistant professor of economics at UC Berkeley, the country is hard to miss in the financial world. "Luxembourg has private banks like Switzerland, it has a big mutual fund industry like Ireland's, it's used for corporate tax avoidance like Bermuda or the Netherlands, and it also hosts one of the two international central depositories for securities, so it's active in euro bonds," he says. "It's the tax haven of tax havens, present at all stages of the financial industry." Tony Norfield, a former banker in the City of London who now writes on global finance, has described Luxembourg as "a paragon of parasitism".

The story of how a marginal and relatively powerless country has survived world wars, economic crises and cataclysmic technological advances to become a banking and finance powerhouse tells us a lot about how far a small country can go if it devotes itself to anticipating and accommodating the needs of global capital. It's a contentious business: for every happy shareholder praising Luxembourg's business-friendly rules and money-saving loopholes, there's a critic condemning Luxembourg's willingness to expedite the regulatory "race to the bottom".

Then again, there aren't many options for a country like Luxembourg besides exploiting its most valuable resource: its national sovereignty. And Luxembourg has done this more and better than any other country in the world. By crafting innovative rules, laws and regulations that only it could (or would) put on offer, Luxembourg has attracted banks, telecommunications companies and consulting firms before any of these industries came to dominate the global economy. Now, by courting asteroid miners before anyone else takes them seriously, it may very well end up doing the same thing for the commercialization of space.

Luxembourg's first significant attempts at liberalization began in the late 1920s and early 1930s. As radio grew popular, the grand duchy decided not to create a publicly funded radio service like its neighbors. Instead, it handed its airwaves to a private, commercial broadcasting company. That company – now known as RTL – became the first ad-supported commercial station to broadcast music, culture and entertainment programs across Europe in multiple languages. "By handing the rights to a public good to a private company, the state commercialized, for the first time, its sovereign rights in a media context," notes a 2000 book on Luxembourg's economic history. The title of the book, published by a Luxembourgish bank, is, tellingly, *The Fruits of National Sovereignty*.

Then, just three months before the stock market collapsed in 1929, Luxembourg's parliament passed legislation exempting holding companies – that is, parent firms that exist solely to own parts of or control other companies – from paying corporation taxes. In the first five years after the law's passing, 700 holding companies were established; in 1960, there were 1,200, and by the turn of the century, some 15,000 "letterbox" firms – one for every 18 citizens – were incorporated in Luxembourg. (In 2006, the European commission found that this exemption violated EU rules, so Luxembourg promptly created a new designation, the "family estate management company", that complied with the country's EU treaty obligations while offering many of the same money-saving advantages.)

Throughout the first half of 20th century, Luxembourg's main industry was steel, but by 1980, that business all but collapsed. Even before its iron ore mines shut down, though, the grand duchy came to represent a discreet but powerful regulatory freedom. A homegrown economic model began to take shape: over the next decades, it would make a name for itself by passing legislation "designed to tempt the world's hot money," notes the Tax Justice Network, an anti-tax-evasion advocacy group.

The country's policymakers also realized that less could really be more. According to Georges Schmit, a lifelong civil servant who has played a big role in shaping the country's economy since he joined the ministry of the economy in 1981, a key component of Luxembourg's early success was the fact that it did not have its own central bank. The country had been in a monetary union with Belgium since 1921, and didn't impose reserve requirements on financial firms. This meant banks could lend or spend the money that they would have had to keep on deposit in other jurisdictions. In Schmit's words,

Luxembourg's biggest draw "wasn't our doing; it was the lack of our doing anything".

Over the years, the government managed to coax over foreign financial institutions, from complex securitization vehicles to Islamic banks. And on the consumer level, the state's low taxes drew Europe's tax-averse petty bourgeoisie. Starting in the 1960s, "Belgian dentists" and "German butchers" – the prevailing stereotypes cited in the international financial press – began taking daytrips to the grand duchy to deposit money to avoid tax at home. The Luxembourgish state even lowered fuel costs to attract the daytrippers, and in 1981, introduced legally binding bank secrecy comparable to Switzerland's.

In the next century, the dentists would give way to Qatari princes, Chinese princelings and other global members of the global super-rich – or at the very least, their investments. "When a country is small, the rest of the world is big," says Schmit. "Since independence we needed to find larger economic spaces, be they regional or continental." By serving as a hub for investors, companies and markets during decades of rapid deregulation and globalization, Luxembourg turned itself into an indispensable cog in the machinery of international finance.

In 2009, Schmit embarked for California to continue his life's work: finding new ways for his country to attract money, this time as the general consul and trade envoy in Silicon Valley.

Since he had joined the ministry of the economy to devise new innovation strategies almost three decades earlier, his country seemed to have defied all odds and made virtues of its apparent weaknesses. Its small size had not prevented it from becoming the largest centre for investment funds in the world after the US. Its tiny population had not deterred multinationals and EU institutions such as the court of justice from basing their headquarters there. It had parlayed its status as a neutral country and founding member of many European organisations into sending three of its politicians – more than any other country – to preside over the European commission. And by marketing its easy access to Europe, an educated workforce, bank secrecy (which it voted to end in 2014 under pressure from other countries and the OECD) and myriad regulatory advantages, the country built an outsized financial sector.

Crucially, Luxembourg never seemed to let an opportunity pass it by. Following its support for commercial radio 50 years prior, the country was the first in Europe to privatise satellite television. In 1985, the grand duchy granted a company called Société Européenne des Satellites (SES) the right to broadcast TV directly to viewers' homes from a satellite positioned in space. "The big innovation is that this was a privatization of space," says Schmit, who served for 17 years on the SES board. "All the other operators were owned by governments through international agreements. This was the first commercial company that set out to use space for broadcasting." When SES grew profitable, Luxembourg's bet paid off: the tiny country became home to a telecoms giant, and, as an early investor, received a piece of the pie.

In the early 2000s, Luxembourg pounced at the chance to court retailers such as Amazon and Apple with tax incentives. There were the perks the state was happy to publicize – the lowest VAT in Europe, for instance – and there were case-by-case deals with large companies that it kept rather quieter. The companies flocked in, but in the aftermath of the financial crisis, with awareness of wealth inequality growing and austerity measures bruising ordinary Europeans across the continent, Luxembourg could only keep these arrangements under wraps for so long.

In late 2014, the grand duchy went from relative obscurity to complete infamy when the details of these "tax rulings" – versions of which were also carried out by Belgium, Ireland and the Netherlands – were disclosed by the International Consortium of Investigative Journalists. Known as the "Lux leaks", the massive trove of leaked data revealed that, from 2002 to 2010, the country's tax agency approved a series of confidential deals that allowed AIG, Ikea, Deutsche Bank and more than 300 other large firms to save billions of dollars they might have otherwise owed to other countries.

The rulings weren't necessarily illegal, and they weren't unique to Luxembourg, but they did cause a scandal, provoking damning reports in the media, protests around Europe and promises for tighter

"Here was a chance to change
the conversation away from taxes
and towards space; to establish
an industry for Luxembourg's future;
to contribute to science and
human knowledge."

regulation from within the EU. Investigations on both sides of the Atlantic on related matters followed, and lawsuits revealed information on more companies still. (One memorable detail: Amazon's 28-step tax-restructuring arrangement in Luxembourg was named Project Goldcrest after the country's national bird.)

Around this time, Zucman, a recent Paris School of Economics PhD who studied with Thomas Piketty, began looking into Luxembourg's role in international tax avoidance and evasion. His focus was not on the multinationals, but on Luxembourg's thriving fund industry, which through niche regulations and loopholes allowed investors to avoid certain taxes, too. Luxembourg was a well-known financial centre, but the statistics Zucman dug up while researching his book, The Hidden Wealth of Nations, took him aback: in 2015, national data showed $3.5tn worth of shares in Luxembourgish mutual funds were domiciled in the grand duchy, while data from other countries accounted for only two of those trillions. The missing $1.5tn suggested to him that the money – which, he notes, was probably accumulating interest by the day – had no identifiable owner. That meant the countries to whom tax was owed on these ungodly sums were unaware of their existence.

Globally, Zucman calculated almost $8tn in financial wealth – which does not include real estate, luxury goods, gold or other commodities – has been stolen from countries and taxpayers in this fashion thanks to "secrecy jurisdictions" such as Luxembourg, the Virgin Islands or Panama working "in symbiosis". In his book, Zucman described Luxembourg as an "economic colony of the international financial industry" and challenged its right to its greatest asset: its sovereignty.

"Imagine an ocean platform where the inhabitants would meet during the day to produce and trade, free of any law or any tax, before being teleported in the evening back home to their families on the mainland," he wrote, referring to the country's unusual demographics: 47% of Luxembourg's 500,000 residents are foreign, and 44% of the workforce commutes in across nation-state lines each day for work. "No one would dream of considering such a place, where 100% of its production is sent abroad, as a nation.

"The trade of sovereignty knows no limits," Zucman continues. "Everything is bought; everything is negotiable. Luxembourg is not the only country that has sold its sovereignty, far from it ... but it is the one that has gone the furthest."

Scrutiny of Luxembourg's tax practices – from the press, the public and the EU – spread at an awkward time. At the end of 2013, the country elected a new prime minister, Xavier Bettel, whose coalition government of democrats, socialists and greens wanted to distance themselves from the economic policies of former prime minister Jean-Claude Juncker and play by the EU's rules. "Honestly, I am fed up with being accused of being a defender of a tax haven and a hotbed of sin," Bettel said in a speech to the Luxembourg Bankers' Association shortly after taking office. "We need to work on our image ... we have much changed in the last years, now it is time to make sure that everybody knows."

Etienne Schneider, then economy minister, was part of this effort, too. But instead of being applauded for breaking with the past, from the moment they took power the politicians were constantly reminded of their country's indiscretions. The new government needed to square the Luxembourgish model of economic development with new political realities. It had to keep looking ahead. Most of all, it wanted to change the conversation.

A curious possibility had emerged the previous summer, when Georges Schmit visited Nasa's Ames research centre in Palo Alto and found himself in conversation with Pete Worden, a former director of the centre. Over coffee, Worden told Schmit about the emerging NewSpace sector and about his dream of finding life on other stars and planets.

Schmit sensed Worden would hit it off with Schneider, so he introduced them. At first, asteroid mining struck Schneider as crazy. "I listened to him and wondered what this guy might have smoked this morning; it sounded like complete science fiction," he recalled. But the more he listened, the more it made sense. Worden persuaded Schneider that "it's not if it will happen, it's when it'll happen. And the countries who'll be the pioneers will be the ones that'll get the most out of it later on."

From 2014 to 2016, a series of meetings between the Americans and the Luxembourgeois took place. If they resembled April's trade mission, they will have involved tedious tours of technology companies, self-aggrandizing speeches about how space would bring about Earth's "third industrial revolution", and many hours stuck in traffic – but also genuine wonder at what might happen if humankind made space their own.

Schneider hung his hopes – and his political future – on the stars. Here was a chance to change the conversation away from taxes and towards space; to establish an industry for Luxembourg's future; to contribute to science and human knowledge, even. Besides, in such trying times, who didn't like talking about the wonders of exploring the great unknown? NewSpace companies were certainly eager to work with Luxembourg. They were thirsty for funds and attention, and felt invisible in the US. Luxembourg was a place where they could get meetings with high-level politicians in minutes; where everyone spoke great English; where the bureaucracy was minimal, and the promise of low taxes remained. As one NewSpace executive told me this year: "We just want to work with a government who won't get in the way."

The only catch was the ambiguity of space law: companies wanted assurances that the fruits of their extraterrestrial labour would be recognised here on Earth. This is not a given. Unlike on Earth, where a country can grant a company a mining concession, or a person can sell the right to exploit their land, no one has an obvious legal claim to what's outside our atmosphere. In fact, the Outer Space Treaty, signed by 107 countries at the UN in 1967, explicitly prohibits countries from claiming sovereignty over celestial bodies. The question now is: if nobody owns or governs the great unknown, who is to say who gets to own a little piece of it?

Since the emergence of the NewSpace sector, individual countries have attempted to lend some clarity to eager entrepreneurs, reasoning that the prospect of private property in space will encourage hard work and innovation. The American Space Act, passed in 2015, is the first "finders, keepers" law that recognizes ownership of space resources, but it only does so for companies owned by US citizens. In October 2015, Luxembourg commissioned a study on whether it could fill that legal void. The report, completed in 2016, noted that "while legal uncertainty remains, under the current legal and regulatory framework, space mining activities are (at least) not prohibited" and concluded that Luxembourg should pass legislation that gives miners the right to keep the extraterrestrial bounty they extract.

Such a law was drafted shortly after the study's completion, and on 1 August 2017, it went into effect. Luxembourg's bill does not discriminate by nationality, or even by the location of a company's headquarters. In fact, the law indicates the country's willingness to serve as a sort of flag of convenience for spacecrafts, allowing them to play by one country's futuristic rules in the absence of universal, binding agreements. Rick Tumlinson, of Deep Space Industries, another space exploration company in which Luxembourg has invested, told me that there was value in Luxembourg's law because it saw no citizens and no borders: just one blue planet from high above.

Six weeks after the trade mission in California, I disembarked from a tiny plane on the runway of Luxembourg City's airport in a melee of grey suits and black carry-on roller bags. I walked past large wealth-management and equity-fund advertisements into the car park, where I caught the bus into the city center, passing dozens of huge new building projects, a tramline under construction and two enormous yellow towers that, in the afternoon light, resembled twin gold bars reaching for the sky.

Within an hour, I was sitting at a table outside a dive bar opposite the old city's bathhouse with Lars Schmitz, 29, and Gabrielle Taillefert, 21, two members of a local theatre and art collective called Richtung22 (Direction22). Over the past few years, the group has staged a series of performances lampooning their country's mercenary modus operandi. Instead of writing their own scripts from scratch, the collective makes dramatic collages almost entirely out of primary documents: laws, press releases, speeches, transcripts from parliament, promotional videos and so on.

One of Richtung22's early works satirized Luxembourg's nation branding committee, which was set up in March 2013 to promote the country abroad. The play, which was financed in part by the culture ministry, was entitled Lëtzebuerg, du hannerhältegt Stéck Schäiss (Luxembourg, Vicious Pile of Shit). Since then, Schmitz says, state funds for Richtung22's work have dried up.

In his spare time, Schmitz, who is slight of build with cropped blonde hair, works on antifascist and anti-capitalist organizing. He has the droll resignation of a leftwing activist operating in a country whose politics are so abstract and so global that grassroots resistance must necessarily come in the form of farce. Richtung22's latest play savages the country's efforts to attract the NewSpace industry. Its title is Luxembourg's Private Space Explorevolutionary Superfancy Asteroid Tailoring. Schmitz sees space mining as a high-tech spin on an age-old scam: selling sovereignty. "The country's business model is hidden," he said. "It's making laws that companies want, and taking a risk on those companies. But the government uses it to say 'This is how modern we are! This is something new!'"

Zucman shares Schmitz's view. "Adapting this strategy to the business of space conquest is what being an offshore financial center means," he says. "It's not diversification. It's just extending the logic of being a tax haven to new area."

On stage, the entire space enterprise is portrayed as a cynical, money-grubbing, reputation-redeeming debacle dictated by private-sector interests. "We feel bad that our country does this to the world, and no one else here talks about this stuff," Schmitz told me. He ran off a dozen or so Luxembourgish transgressions, including but not limited to aiding and abetting tax evasion and weaseling its way out of EU banking regulations. In such a small country, it's hard to be so outspoken against the national interest. "People think we're traitors," he said.

Was there anything good about his country, I asked. "It's beautiful," Schmitz conceded. He was right: Luxembourg is beautiful, and was particularly charming on that balmy May evening. The city rests on two levels; the smaller "low" city's quaint little streets and sidewalk cafes skim the river, while the "high" city center is home to a lively main drag with pricey boutiques, fancy chocolate shops and chains such as H&M. Cafes advertise crémant – a local bubbly wine – and local dishes that borrow their richness from the French and their stodginess from the Germans.

The next day, I went to meet Marc Baum, an MP from the democratic socialist party déi Lénk (the Left). He handed me a policy paper his party published criticizing Schneider's space-mining proposal: they believe his law is inconsistent with Luxembourg's outer-space treaty obligations, that it creates opportunities for billionaires to further enrich themselves and could be harmful to the environment. Even worse, it enshrines the notion of "competition instead of cooperation" between states. "It's infinite capitalism!" Baum exclaimed over a cold beer on a terrace.

Baum, as it happened, is an actor, too. When we met, he was preparing to perform Eugène Ionesco's Rhinoceros, an absurdist play about a town whose protagonists speak exclusively in clichés and end up turning into rhinos on account of their unquestioning conformity. Over the course of the drama, the townspeople justify their decision to "go rhino" by declaring that "humanism is dead, those who follow it are just old sentimentalists". The play's sole hero, Berenger, resists succumbing to "rhinoceritis", but fails to save anyone else: he ends up being the only person in the whole town who does not grow a horn. The analogy between that and Baum's own predicament seems a little on the nose. He was one of just two politicians who voted against the space law in July.

In June, about a month before his signature legislation was passed by the parliament, Schneider and some of his associates flew to New York for yet another sales pitch – this time, for the benefit of venture capitalists on the east coast. His speech focused on the financial aspects of Luxembourg's space race, and the country's intention to get in on the ground floor of commercial space exploration. "Under the US Space Act, your capital has to be majority US capital," he said, referring to US willingness to recognize property rights in space for its citizens. "We don't really care where the money comes from in our country, as long as the money is clean."

On Schneider's telling, Luxembourg could do for the space-resource trade what it had done for the eurodollar market, international holding companies and multinationals: provide a safe, reliable base where they could operate in tandem with a keen and cooperative – or, by his detractors' assessment, pliable and sycophantic – state. Schneider announced that after passing its law, Luxembourg would create its own space agency. This would not be a copy of Nasa, but would instead "focus only on commercial space resources". He told the audience that Luxembourg would solicit private funding to capitalize NewSpace companies, and seek the advice of venture capitalists to decide what companies to invest in. If asteroid mining does, in fact, take off, Luxembourg will be what Schneider's friends in Silicon Valley might call an "early adopter".

It's a gamble, for sure. But it's difficult to imagine where Luxembourg would be had it not deployed this ingenious development strategy continuously over the past century. The global economy offers few alternatives than to serve it, and rewards its enablers richly. Perhaps a mercenary spirit is what it takes to succeed as a small country in the world – and that "we want to remain what we are" is just Luxembourgeois for the old French saying: *plus ça change*.

What Will Kill Neoliberalism? A Roundtable on Its Fate

By Joelle Gamble,
Paul Mason, Bryce Covert,
William Darity Jr.
& Peter Barnes

(Originally appeared
in the *Nation*, May 2017)

*M*assive global inequality underlies our era of economic and political unrest. The rise of nationalist, populist movements, and the faltering influence of the Davos class of free-trade advocates, have rendered neoliberalism an ideology without committed ideologues. So what will bring about the end of neoliberalism—the left? the right? the incompetence of the professional political class?—and, when it's gone, what will replace it? The Nation asked five of its favorite minds for their views on the direction we urgently need to go next.

POPULISM ASCENDANT (Joelle Gamble)

The dramatic effects of deindustrialization, automation, globalization, and the growing disparities of wealth and income—including by race and region—are undermining political norms in much of the West.

Activists and academics alike have linked these trends to the neoliberal ideology that has guided policy-making over the past several decades. This ideology has resulted in pushing the widespread deregulation of key industries, attempting to solve most social and economic problems through market competition, and privatizing public functions like the operation of prisons and institutions of higher education. Neoliberal ideas were considered such common sense during the 1980s and '90s that they were simply never acknowledged as an ideology. Now, even economists at the International Monetary Fund are willing to poke holes in the ideology of neoliberalism. Jonathan Ostry, Prakash Loungani, and Davide Furceri wrote in 2016: "The costs in terms of increased inequality are prominent. Such costs epitomize the trade-off between the growth and equity effects of some aspects of the neoliberal agenda."

We know that neoliberalism has now provoked populist responses on the left and the right. But are either of them sufficient to end its rule?

Left populism, if organized, could end the neoliberal order: As espoused by leaders like Pramila Jayapal and Keith Ellison, left populism demands public control as well as redistribution; it is pro-regulation, pro-state, and anti-privatization. These values are inherently at odds with the small-government, anti-regulatory tenets of neoliberalism. If an aggressive left-populist agenda is successfully implemented, neoliberalism would be defeated. The barrier to implementation is the left's inability to be consistent and organized.

Populism on both the left and right has proved difficult to organize and suffers from a lack of leadership. On the left, the struggle for organization has been playing out in the Democratic Party's leadership fights. Politicians and activists are attempting to close the ideological gap between the party's base and its leaders. Without enough trust to allow leaders to set and execute a well-resourced strategy—to say nothing of the resources themselves—the left faces huge obstacles to actually implementing an agenda that spells the end of neoliberal dominance, despite having an ideology that could usher in a post-neoliberal world.

Left populism can technically end neoliberalism. But can right-wing populism?

One should hope that right-wing populism doesn't become organized enough to end the neoliberal order. Public control is not a cogent ideology on the right. That leaves room for privatization—a main pillar of neoliberalism—to continue to grow. Only if right-wing nationalism turns into radical authoritarian nationalism (read: fascism) will its relationship with corporate power turn into an end to the neoliberal order. In the United States, this would mean: 1) the delegitimization of Congress and the judicial branch, 2) the increased criminalization of activists and political opponents, and 3) the nationalization of major industries.

Right-wing nationalism seems to be crafted to win electoral victories at the intersection of protectionist and xenophobic sentiments. Its current manifestation, designed to win over rural nativist voters, appears to be at odds with the pro-free-trade policies of neoliberalism. However, the lines between far-right nationalism and the mainstream right are blurring, especially when it comes to privatization and the role of government. In the United States, Trump's agenda looks more like crony capitalism than a consistent turn from neoliberal norms. His administration seems either unwilling or incapable of taking a heavy-handed approach to industry.

As with many of his business ventures, we've already seen Trump-style nationalism fail in his nascent administration. The White House caved to elite Republican interests with the attempt to repeal and replace the Affordable Care Act and with Trump's decision to stack high-level economic-policy roles with members of the financial elite. Trump's proclaimed nationalist ideology seems to be a rhetorical device rather than a consistent governing principle. It's possible that the same might be true for other right-wing nationalists. France's Marine Le Pen has cozied up, though admittedly inconsistently, to business interests; she has also toned down her rhetoric, especially on immigration, over the years in order to win centrist voters. Meanwhile, Dutch nationalist Geert Wilders notably lost to a more mainstream candidate in March's general elections. Yet the radical right is more organized in Europe than in the United States. We may not see the same level of compromise and incompetence as in the Trump administration. Moves toward moderation may only be anomalous and strategic rather than a sign of a failing movement.

So what does all of this mean for the future of neoliberalism, particularly in the American context? I believe there are two futures in which neoliberalism's end is possible. In the first, the left decides to stop playing defense and organizes with the resources needed to build sustained power, breaking down the policies that perpetuate American neoliberalism. This means enacting policies like universal health care and free college, and ousting the private-prison industry from the justice system. In the second future, a set of political leaders who have been emboldened by Trump's campaign strategy gain office through mostly republican means. They could concentrate power in the executive in an organized manner, nationalize industries, and criminalize communities who don't support their jingoistic vision. We should hope for the first future, as unlikely as it seems in this political moment. We've already seen the second in 20th-century Europe and Latin America. We cannot live that context again.

TAKE THE STATE (Paul Mason)

I wrote in *Postcapitalism: A Guide to Our Future* that if we didn't ditch neoliberalism, globalization would fall apart—but I had no idea that it would happen so quickly. In hindsight, the problem is that you can put an economy on life support, but not an ideology.

After the 2008 financial crisis, quantitative easing and state support for banks kept the patient alive. As the Bank of England governor Mark Carney said last year at the G20 summit in Shanghai, central banks have even more ammunition to draw on should they need it—for example, the extreme option of "helicopter money," in which they credit every bank account with, say, $20,000. So they can stave off complete stagnation for a long time. But patchwork measures cannot kick-start a new era of dynamism for capitalism, much less faith in its goodness.

The human brain demands coherence—and a certain amount of optimism. The neoliberal story became incoherent the moment the state had to take dramatic steps to support a failing financial market. The form of recovery stimulated by quantitative easing boosted the asset wealth of the rich but not the income of the average worker—and rising costs for health care, education, and pension provision across the developed world meant that many people experienced the "recovery" as a household recession.

So they began looking for answers, and the right had an easy one: Ditch globalization, free trade, and relatively free migration rules, as well as acceptance of the undocumented migrants who keep the economy working. That's how we get to Donald Trump, Marine Le Pen, Geert Wilders, Viktor Orbán, the Law and Justice party in Poland, and UKIP in Britain. Each of them has promised to make their country "great again"—by diverting growth toward it and migrants and refugees away.

For 30 years, neoliberalism taught national elites that they were better off collaborating in the creation of a positive-sum game: Everybody wins, ultimately, even if your factory moves to China. That was the rationale.

Economic nationalism is logical if you believe that stagnation will last a long time, creating a zero-sum or even a negative-sum game.

But the projects of economic nationalism will fail. This is not because economic nationalism has always been a losing strategy: Adolf Hitler practically abolished German unemployment within five years, and Franklin Roosevelt triggered a spectacular recovery and reindustrialization with the New Deal. But these were programs of another era, in which business models were primarily national and monopolies operated in the sphere of one big nation and its colonies; where the state was heavily enmeshed in the national economy; and where global trade was puny and economic migration low compared to now.

To try a repeat of autarky in the 21st century will trigger dislocation on a large scale. Some countries will win: It's even feasible that, although led by an imbecile, the United States could win. However, "winning" in this context means bankrupting other countries. Given the complexity and fragility of the globalized system, the cities of the losing nations would resemble New Orleans after Hurricane Katrina.

In the long term, for the left, the transition to a system beyond capitalism must be based on the possibility of a low-work, high-abundance society. This is the essence of the postcapitalism project that I proposed: automate work, replace wages with a basic income and heavy state provision of services, and enforce competition among the rent-seeking monopolies in order to force the price of their goods so low that people can survive scarce and precarious work.

As Manuel Castells's research group in Barcelona has found, as the market staggers, more and more people actually begin to adopt non-market survival tactics, mechanisms, and institutions like informal lending, co-ops, time banks, and alternative currencies. And that's the basis for an economic counterpower to big capital and high finance.

But in the short term, a whole generation of the left that reveled in aimlessness and horizontality needs to split the difference between that and effective, organized politics. Call it "diagonality," if you want: Without ceasing to care about the 100 small causes that have animated us in the past, the one big cause that needs to animate us in the future is a systemic project of transition beyond capitalism. For now, that project has to be pursued at the level of big cities, regions, states, and alliances of states—that is, at scale.

The hardest thing for the old left to accept will be that this means using the existing, oppressive, imperfect state while simultaneously trying to democratize it. Street protests, mass resistance, strikes, and the occupation of squares are great ways to assemble the forces. But the arc of the story from 2011 to 2015—Occupy, the Indignados, and the Arab Spring—shows that we have to do more than simply create a counterpower: We need to take power and diffuse it at the same time.

THE CRISIS OF CARE (Bryce Covert)

American parents are being crushed between trying to care for their families and working enough hours to survive financially. This problem plagues parents of both genders, up and down the income scale, and it is upending the way Americans view the capitalist system. This crisis of care is fostering solidarity among the millions of Americans who share this challenge, as well as support for solutions that will end the reign of neoliberalism.

Among low-income Americans, especially people of color, both parents have often worked outside the home to make ends meet. Nonetheless, the ideal has been, until very recently, a stay-at-home mother and a father working for pay outside the home. World War II undermined this idyll, pushing women into factories as men went to fight abroad. The gauzy 1950s dream of single-earner families masked the reality that women continued to pour into the workforce.

Today, women make up about half of the paid labor force in the United States, including more than 70 percent of women with children. This means that in about half of married heterosexual couples, both the husband and wife work. This has given women far more access to the public sphere and, with it, greater status and equality both inside and outside the home.

But it's also meant a crunch for families. There is no longer a designated parent to stay home with the kids or care for aging relatives, and the workplace isn't designed to help with that predicament. Instead, work is devouring people's lives.

You can see this problem in the rising number of Americans who worry about their work/life balance. About half of parents of both genders say they struggle to reconcile these competing demands. Fathers are particularly freaked out: More than 45 percent feel they don't spend enough time with their children, compared with less than a quarter of mothers (probably because more women reduce their paid work to care for children). As the baby-boomer generation ages, a growing elderly population threatens to trap even more working people in the predicament of caring for aging parents, raising young kids, and trying to make a living.

The result has been that more and more people are being forced to reckon with the fact that capitalism's unquenchable thirst for labor makes a balanced life impossible. This, in turn, is fostering a greater sense of solidarity among them as workers struggling against the demands of corporate bosses. This growing crisis has already led to some policy-making. The expansion of overtime coverage by the Obama administration means that workers will either be better compensated for putting in long hours or have their schedules pared back to a more humane 40-hour work week (though it remains to be seen what will happen to the overtime expansion under President Trump). Legislation guaranteeing paid time off has swept city and state governments. These are policies that challenge the idea that we should give everything of ourselves to our jobs.

The crisis of care has also revived the notion that the public should deal with these shared problems collectively. While other developed countries have spent money to create government-funded solutions for child care over the past half century, Americans have insisted child care remain a private crisis that each family has to solve alone. The United States provides all children age 6 to 18 with a public education, but for children under the age of 6, it offers basically nothing. Head Start is available to some low-income parents, and a smattering of places have started experimenting with universal preschool for children ages 3 and 4. Outside of that, parents are left to a pitiful private system that often doesn't even offer them enough slots, let alone quality affordable care.

Americans have increasingly come to recognize that this situation is ridiculous and are throwing their support behind a government solution. Huge majorities support spending more money on early-

"American parents haven't yet gone on strike against capitalism's endless demands on their time or the government's failure to provide public support. But the crisis is reaching a boiling point..."

childhood programs. American parents haven't yet gone on strike against capitalism's endless demands on their time or the government's failure to provide public support. But the crisis is reaching a boiling point, and it's transforming our relationship to America's neoliberal system.

A REVOLUTION OF MANAGERS (William Darity Jr.)

Marx's classic law of motion for bourgeois society—the tendency of the rate of profit to fall—was the foundation for his prediction that capitalism would die under the stress of its own contradictions. But even Marx's left-wing sympathizers, who see the dominant presence of corporate capital in all aspects of their lives, have argued that Marx's prediction was wrong. It has become virtually a reflex to assert that modern societies all fall under the sway of "global capitalism," and that a binary operates with two great social classes standing in fundamental opposition to each other: capital and labor.

Suppose, however, that Marx was correct in his expectation that capitalism, like other social modes of production before it, will wind down gradually, but wrong in his expectation that it would be succeeded by a "dictatorship of the proletariat," a civilization without class stratification. Suppose, indeed, that the age of capitalism is actually reaching its conclusion—but one that doesn't involve the ascension of the working class. Suppose, instead, that we consider the existence of a third great social class vying with the other two for social dominance: what was seen in the work of such disparate thinkers as James Burnham, Alvin Gouldner, Barbara Ehrenreich, and John Ehrenreich as the managerial class.

The managerial class comprises the intelligentsia and intellectuals, artists and artisans, as well as state bureaucrats—a credentialed or portfolio-rich cultural aristocracy. While the human agents of global capital are the corporate magnates, and the working class is the productive labor—labor that is directly utilized to generate profit—the managerial class engages comprehensively in a social-management function. The rise of the managerial class is the rise to dominance of unproductive labor—labor that can be socially valuable but is not a direct source of profit.

A surplus population under capitalism has a purpose: It exists as a reserve army of the unemployed, which can be mobilized rapidly in periods of economic expansion and as a source of downward pressure on the demands for compensation and safe work conditions made by the employed. Therefore, capital has little incentive to eliminate this surplus population. In contrast, the managerial class will view those identified as surplus people as truly superfluous. The social managers consider population generally as an object of control, reduction, and demographic administration, and whoever is assigned to the "surplus" category bears the weight of the arbitrary.

To the extent that identification of the surplus population is racialized, particular groups will be targets for social warehousing and extermination. The disproportionate overincarceration of black people in the United States—a form of social warehousing—is a direct expression of the managerial class's preferences regarding who should be deemed of low necessity. The exterminative impulse is evident in the comparative devaluation of black lives that prompted resistance efforts like the Black Lives Matter movement. The potential for black superfluity in the managerial age is evident in prescient works like Sidney Willhelm's *Who Needs the Negro?* (1970) and Samuel Yette's *The Choice* (1971), both published almost 50 years ago.

The assault on "big" and invasive government constitutes an attack on the managerial class by both capital and the working class. Despite endorsing military spending, receiving lucrative government contracts, and enjoying the benefits of publicly provided infrastructure like roads, highways, and railways, corporate capital calls for small government. This is a strategic route to slashing social-welfare expenditures, with the goal of reducing the wage standard and eliminating all regulations on corporate predations. Despite benefiting from social-welfare expenditures, the working class gravitates to a new brand of populism that blends anticorporatism with anti-elitism (and anti-intellectualism), xenophobia, and a demand for a smaller and less intrusive state. Since "big" government constitutes the avenue for independent action on the part of the managerial class, an offensive of this type directly undermines the "new" class's base of power.

But the managerial class also possesses another attribute that is both a strength and a weakness. Unlike capital and labor, whose agendas are driven to a large degree by the struggle over the character of a society structured for the pursuit of profit, the managerial class has no anchor for its ideological stance. In fact, it's a social class that is wholly fluid ideologically. Some of its members align fully with the corporate establishment; indeed, the corporate magnates—especially investment bankers—look much the same as members of the managerial class in terms of educational credentials, cultural interests, and style. Other social managers take a more centrist posture harking back to their origins in the "middle class," while still others position themselves as allies of the working class. And there are many variations on these themes.

Depending on where the ideological weight centers most heavily, the managerial class can take many directions. During the wars in southern Africa against Portuguese rule, Amílcar Cabral once observed that for the anticolonial revolution to succeed, "the petty bourgeoisie" would need to commit suicide as a social class, ceasing their efforts to pursue their particular interests and positioning themselves fully at the service of the working class. One might anticipate that the global managerial class will one day be confronted with the choice of committing suicide, in Cabral's sense, as a class. But the question is: If such a step is taken, will they place themselves fully at the service of labor… or capital?

UNIVERSAL BASE INCOME (Peter Barnes)

There is no single solution to economic inequality and insecurity in America, but there's one that could go further than any other. It's a universal base income, as distinct from a universal basic income.

A universal base income of a few hundred dollars a month is not the same as a universal basic income of, say, $1,000 a month. The latter, at least in some places, is enough to survive on; the former decidedly is not. And while the latter is the dream of many, it is far too expensive—and threatening to America's work ethic—to be enacted anytime soon. If a universal basic income ever happens here, it will be because it was preceded for many years by a universal base income, gradually nudged upward like Social Security and the minimum wage. So let's take a look at that.

A universal base income is both a springboard and a cushion for every participant in our fast-changing market economy—like giving everyone $200 for passing "Go" in a game of Monopoly. It supplements, but does not replace, labor income (which for the last 30 years has stagnated or declined), and it does so without judgment or stigma. It is grounded on the principle that, in a prosperous albeit volatile and increasingly unequal economy, everyone has a right to some cash flow they can count on.

In practical terms, a universal base income would be simple to administer. Eligible recipients (anyone with a valid Social Security number, which can include legal immigrants) would receive an equal amount of money every month, wired to their bank accounts or debit cards. The system would look and feel like Social Security, or a monthly version of the dividends that all Alaskans receive. People who don't need the extra income would be enabled by a check-off option to donate it to any IRS-approved charity.

A universal base income, I should note, has nothing to do with automation, robots, or artificial intelligence. It has a lot to do with enhancing every American's security, reducing their stress, and giving our poor and middle classes a leg to stand on—the very opposite of what our economy does now.

A universal base income would have other benefits as well. It is an answer—perhaps *the* answer—to long-term economic stagnation, a trickle-up form of Keynesianism that would stimulate our economy through increased household spending. Moreover, if funded by fees on unproductive activities like pollution and speculation, it would help solve two other deep problems of 21st-century capitalism: climate change and financial instability. And it wouldn't need to replace or reduce spending on current programs that benefit the poor, a regressive trade-off that conservatives favor but most progressives oppose.

There are six large demographic groups (with some overlap) that could form the core of a movement for a universal base income: millennials (the first generation of Americans destined to earn less than their parents), low-wage and on-demand workers (the so-called precariat), women (who still earn less than men and aren't paid at all for much of the work they do), African Americans (who suffer from past and present injustices), retired and near-retired workers (who can't live on Social Security alone), and poor people of all colors. Environmentalists might also link arms with the cause if one of the revenue sources is a tax on pollution. It will, of course, be no simple feat to persuade these diverse groups that what they can't achieve separately they may be able to achieve together. But it has happened before, and, in the post-Sanders era, it could happen again.

In the political realm, a universal base income would bring our nation together by affirming that we are all in the same economic boat. It would unite our desperate poor and our anxious middle class, young and old, women and men, white people and people of color. It would make millions of Americans less stressed, healthier, and perhaps even happier. And it could make many of us proud to be American.

Fourscore and two years ago, Franklin Roosevelt's Committee on Economic Security produced the classic report that led to passage of the first Social Security Act. The report itself went beyond security for the aged. It proclaimed: "The one almost all-embracing measure of security is an assured income. A program of economic security, as we vision it, must have as its primary aim the assurance of an adequate income to each human being in childhood, youth, middle age, or old age—in sickness or in health."

The committee added that, for reasons of political expediency, it was proposing only an assured income for the elderly, but it hoped that the rest of its vision would be implemented in the not-too-distant future. Much of it has been, but not all. A lifelong base income, along with health insurance for all, are the next pieces.

Magic Bullets

By Patrick Blanchfield

(Originally appeared in
Logic Issue 3: Justice,
December 2017)

O n May 2, 2013, in a dusty outdoor firing range somewhere in Texas, a lanky young man with grand ambitions fired what he hoped would be a shot heard round the world—or at least around YouTube.

With a tug of a cord from a prudent distance away, twenty-five-year old law student Cody Wilson pulled the trigger on a crude gun made of 3D-printed Acrylonitrile Butadiene Styrene (ABS) plastic, successfully discharging a single .380 caliber bullet. The fact that the gun, which Wilson dubbed the "Liberator" (after a much-mythified World War II guerrilla pistol), misfired on a subsequent shot, and promptly exploded when loaded with a more powerful cartridge, seemed unimportant: proof of concept had been achieved.

A savvy entrepreneur and self-promoter, Wilson leveraged his test run of the Liberator to the fullest. Posting CAD files for the gun online via his nonprofit, Defense Distributed, Wilson hyped the ep-ochal, disruptive character of his invention in the breathless profiles of him that appeared in *Wired*, *Forbes*, and *The New Yorker*.

Never shy about his self-styled "techno-anarchist" politics—his other ventures have included forays into crypto-currency and "Ha-treon," a Patreon alternative "absent speech policing"—Wilson articulated his vision of the future, sounding both like a Silicon Valley libertarian and a vintage American reactionary: "I think the future is openness to the point of the eradication of government. The state shouldn't have a monopoly on violence; governments should live in fear of their citizenry." The Liberator might not be much as a weap-on on it own terms, Wilson conceded, but it embodied something much bigger: an inexorable future wherein the "myth" of gun con-trol would be "exploded."

Lawmakers and media were quick to respond to Wilson's proph-ecies with drama of their own. Senator Chuck Schumer from New

"I think the future is openness
to the point of the eradication
of government. The state shouldn't
have a monopoly on violence."

York proclaimed that the advent of 3D-printed firearms meant that, "A terrorist, someone who's mentally ill, a spousal abuser, a felon can essentially open a gun factory in their garage."

His colleague, Congressman Steve Israel, proposed legislation to ban them. "Security checkpoints will do little good if criminals can produce plastic firearms and bring those firearms through metal detectors into secure areas like airports or courthouses," Israel told *Wired*. "When I started talking about the issue of completely plastic firearms, I was told the idea of a plastic gun is science fiction. That science fiction is now a dangerous reality."

False Prophets

Behind all the tumultuous political rhetoric, and behind Wilson's showmanship, there was just one problem: the "3D-printed gun" wasn't technically all 3D-printed. It was hardly "completely plastic" either. The Liberator's crude firing pin—a generic hardware store nail—would absolutely set off metal detectors. And even if it didn't, the metal in the bullets certainly would.

Subsequent, more sophisticated iterations of 3D-printed firearms have yet to overcome this hurdle, and there is no compelling reason to think they will. As far as the specific threat of undetectability is concerned, the hysteria over 3D-printed weapons resembles another panic in the summer of 1990, when a Bruce Willis line in *Die Hard 2* about a (nonexistent) "Glock 7," a "porcelain gun made in Germany ... [that] doesn't show up on your airport X-ray machines" led to terrified public demands for a ban—and, not coincidentally, sold a lot of Glocks.

But panics over new firearms technologies are older than plastic guns or action movies. So are grandiose techno-futurist claims.

The American-born Hiram Maxim, inventor of the first real machine gun, confidently predicted that his creation would actually "make war impossible," rather than producing more lethal conflicts. Never mind that, in more unguarded moments, Maxim would admit that his inspiration to enter the arms industry had come from a businessman friend who had told him, "Hang your chemistry and electricity!

If you want to make a pile of money, invent something that will enable these Europeans to cut each others' throats with greater facility."

In full philosopher-salesman-prophet mode, Maxim insisted that his fearsome weapon would, through a kind of logic of mutually assured destruction *avant la lettre*, leave nations too terrified of mass casualties to ever actually go to war. Needless to say, a brutal century-and-a-half later, Maxim's sales pitch seems either laughably naive or contemptibly cynical.

Evaluating the prophecies of gun futurists, then, the novelty (or lack thereof) of their inventions seems less important than the question of what problems, exactly, they claim to solve. And, by the same token, our collective fascination with gun futurism—our reactions, variously hopeful or hysterical, utopian or bleak—are more interesting when seen in light of what we *don't* find interesting.

It's possible that 3D-printed guns may grow more common, but such fixation on a DIY firearm that requires an $8,000 Stratasys Dimension SST 3D printer to produce seems, at the very least, peculiar in a nation where reliable, durable pocket pistols can be bought for under $100. And our hysteria over the prospect that criminals might order guns via dark-web arms markets, or build assault rifles using homebrew CNC mills in black-market makerspaces, seems likewise misguided given how easily guns can be bought and sold without any paperwork or background checks at gun shows or through private sales in the vast majority of US states.

In other words: the real issue isn't our fantasies about the future, but how focusing on them helps us ignore the legacy of our past and the realities of our present.

Guns Mean Too Much

Firearms occupy a singular place in our national mythology, our legislative landscape, and our political debates. No other object functions as such a ubiquitous icon for key moments in America's past, its different incarnations tied evocatively to various eras—the muskets of Continental soldiers and frontiersmen, the six-shooters of cowboys and desperadoes, the Tommy guns of Chicago gangsters and D-Day paratroopers, the M16s of GIs in Vietnamese jungles, and the AR-15s of open-carry protesters in state capitol buildings.

No other object is addressed so explicitly or at such length in our Constitution, no matter what you might think of either the Second Amendment itself or the convoluted history of its competing interpretations. And no other object is quite comparable as an icon of contested cultural identities and as a flashpoint for vicious partisan disputes.

Cognitive psychologists have documented how guns appear to activate our "affect heuristics": when a senator holds a rifle up for a photo-op on the floor of Congress, or a researcher displays a picture of a gun to a subject in a lab, most people *will* have some kind of immediate and intense reaction, whether positive or negative—a knee-jerk response that belies our ability to dispassionately assess arguments or statistics. Meanwhile, when it comes to the ballot box, attitudes towards gun ownership have arguably become the single biggest predictor for party affiliation and voting preference.

This surplus of meaning—what a psychoanalyst would rightly call an overdetermination—can make debates over guns and gun control both endlessly fascinating and terminally intractable. But it is precisely this overdetermination that occludes the basic realities of political economy that have produced our contemporary situation. These realities have dictated both why and how guns are present in America, and the purposes to which they are put. Partisan polemics and techno-futurist pipe dreams aside, they also represent a hard constraint on the range of possible gun futures.

Strange Bedfellows

It is often observed that Americans own more guns than any other nation. This is true, both per capita and in total numbers. Pinpointing precise figures can get contentious, but well-grounded assessments put the overall number of civilian-owned guns in the United States at well over 310 million, which means there are more guns than there are Americans to own them (112.6 guns for every 100 Americans).

This puts America firmly ahead of its nearest competitors, Serbia (75.6 guns for every 100 Serbians) and Yemen (54.8 guns for every 100 Yemenis). After those top three, the countries with the next highest per-capita civilian gun ownership rates are Switzerland (45.7), Finland (45.3), Cyprus (36.4), Saudi Arabia (35), Iraq (34.2), Uruguay (31.8), and Sweden (31.6). Clustered closely near Sweden, but outside of the global top ten, lie Norway, France, Germany, Canada, Austria, and Germany.

Many commentators will move immediately from these figures to discussions of crime rates and homicide data, or to heavy-handed pontifications, frequently dripping with barely disguised racist exceptionalism, about what (or rather, whom) Americans "should" be like. Bigotry aside, what this move ignores—artfully or naively—is a comparison of civilian rates of gun ownership with those of *non*-civilian gun ownership.

These are striking, since, once again, they reveal that America remains a leader of sorts. When compared to those of the other top fifteen highest civilian-gun-owning nations, the American military arsenal includes over twice as many firearms as its second closest competitor (France), three times as many as its third (Sweden), four times as many as the fourth (Serbia again), and seven times as many as the fifth (Iraq). When it comes to police stockpiles, America once again leads the pack: American police have 1.15 million guns, a number followed only by Iraq (690,000), France (218,000), Yemen (210,000), and Saudi Arabia (90,000).

How to make sense of this—what commonalities shape this distribution? One thing that leaps out, causing considerable offense to chauvinistic American sensibilities, is how, when it comes to being saturated with both civilian guns as well as guns in the hands of its military and police, America far exceeds variously repressive or chaotic Middle Eastern states (Saudi Arabia, Yemen, Iraq) and nations which in recent memory have seen brutal civil wars or other violence (Serbia, Cyprus).

If one expands the set of comparisons on police and military stockpiles to other OECD states, America also starts to resemble our unfortunate next-door neighbor Mexico. There, civilian rates of legal gun ownership are quite low—but Mexico is also where ongoing violence between drug cartels and security forces generates a yearly body count on par with what one would expect from an outright civil war.

But then there are our other peers, in Canada, Western Europe, and the Scandinavian countries. Much beloved as go-to benchmarks of stability and low gun crime for liberal American pundits, these states also have surprisingly large quantities of guns, civilian and otherwise. What deeper structure produces this strange set of statistical bedfellows?

The answer, simply put, is capitalism in general, and the arms trade in particular.

Capitalism Kills

Global military expenditure is estimated to be around $1.6 trillion in any given year. The global trade in weapons clocks in at about $60 billion annually. The United States has dominated this landscape since the 1960s, and in the past two decades has only further cemented its position as far and away the world's largest exporter of arms, with the world's most profitable arms firms.

The global trade in small arms (guns) is, in dollar terms, relatively small: only around $4 billion a year. But its outsize impact is belied by the low cost of individual units—entire army divisions can be equipped with assault rifles for the equivalent cost of a single helicopter or jet—and the fact that, when it comes to casualties, small arms are the leading global killer, far more than missiles, bombs, or tanks.

The list of countries with high gun saturation reflects the dynamics of the global distribution of guns more broadly. Guns from more stable, developed, manufacturing states go to less stable, developing, consumer nations—from West to East, from Global North to Global South. The more contingent cultural features of gun ownership (for example, robust hunting traditions) are merely epiphenomenal to these broader dictates of supply and demand, flows of capital and materiel.

There are the places where guns come from, and where they are plentiful as a matter of course; and then there are the places where guns go, and where they are used to full, lethal effect. These are the two categories of countries with large gun stockpiles: producers and consumers. Supply and demand—it's as simple as that.

Of course, America is, as so often, unique. While other countries like Germany or France manufacture weapons primarily for export, America builds them for both internal and external battlefields—and imports a great many besides. Indeed, the American model of policing unsubtly resembles American practices of foreign military intervention and occupation, and the policies and rhetoric of our security forces reflect this.

Abroad, young brown men killed by US drones are de facto labeled "enemy combatants" by virtue of their age, gender, and location; at home, young black men shot by police are almost invariably said to have a "gang affiliation" based upon similarly gross logics of geographic proximity and networks of family and friends. And there is a sense, too, in which the Bush Doctrine of "anticipatory self-defense" more than passingly resembles the shoot-first attitude of many police departments, where the recourse of "fear for one's life" implicitly structures any encounter between police and civilians.

Homologies aside, the weapons that American troops use in the Global War on Terror have a way of winding up in the hands of domestic police, thanks to initiatives like the 1033 Program, which literally recirculates hardware from US military abroad to security forces back home.

And then there is the unique role and cultural status of the American military proper. As a way of doing business, America's privatized, volunteer military infrastructure contrasts starkly with that of states with mandatory service models like Sweden. As a matter of ethos, the paramilitary overtones of American gun ownership are also unique. In the US, guns are often possessed by individuals as putative tools to be used *against* the state. In countries like Switzerland, private citizens possess guns *on behalf* of the state, and activities like target shooting and institutions like gun clubs receive government subsidies as part of a broader military readiness program.

Indeed, unlike other nations, which could be said to *have* gun cultures, the United States could more accurately be said to be a gun culture. Likewise, unlike any other country on the planet, American politics are shaped by a veritable gun culture *industry*.

As a nation with a heavily privatized military, and where the consumer's ability to buy practically anything is seen as a basic human right, the singular saturation of guns in the American context should actually be fairly unsurprising. And so, too, should the clear disparities in how the toll of gun violence is distributed along uniquely American fractures of race, gender, and class.

Dubious Futures

These are the forces that have helped write the recent American history of guns. Add to this picture the fact that, like few other mass-produced consumer goods, guns are durable, easily concealed, fairly simple to operate, and retain considerable resale value, and we can see how the ubiquity of firearms has vexed pro-gun-control lawmakers from the start.

This obdurate reality goes a long way towards explaining our fascination with "new" guns and gun-related technologies. Faced with a landscape of byzantine regulations, fierce cultural debates, and legislative deadlock, both gun rights advocates and gun control supporters place their hopes in new technology, gambling that some breakthrough might produce a categorical, once-and-for-all break from our murky and ugly present, whether in the direction of absolute gun liberalization or total control.

For gun rights advocates like Cody Wilson, the future is DIY. While major firms focus on other developments—exploring new frontiers in modularity, concealability, subsonic ammunition, suppressor technology, and the legal gray area between rifles and pistols—pro-gun techno-futurists dream of reinventing the supply chain from the ground up. If consumers can make their own firearms, they argue, any gun control laws can be decisively circumvented. Never mind that America's gun control landscape is already a loophole-ridden, contradictory mess, and that cheap, reliable guns are already abundantly available to consumers, legally or otherwise.

On the other side, gun control advocates are also regularly seduced by techno-futurist fantasy. Some hail the advent of "smart guns"—weapons that only fire when wielded by their "proper" owner. At least one such gun—the Armatix iP1, which requires the user to wear an RFID-chip bracelet when shooting—has been available for some time. But canny hackers have already circumvented its mechanism using $15 worth of magnets. And there are other, more fundamental concerns too.

Although there is indeed a market for this kind of weapon (particularly among gun-owning parents), most people who buy guns for self-defense place a premium on reliability and ease of use. A gun that requires the user to first strap on a wristwatch, or that could be susceptible to battery failure or wireless jamming, does not recommend itself as a self-defense weapon. Other putative smart-gun technologies, like fingerprint-enabled triggers and handguards, are also dubious on this front. Reaching for a gun, only to have it not fire because of sweat, rain, or blood, means the weapon is reduced to an expensive club.

More broadly, fantasies about replacing America's massive civilian stockpiles of "dumb" guns with smart alternatives run up against the realities of politics. The political climate remains hostile: a New Jersey law from 2016 mandating that gun dealers eventually transition to smart-gun-only inventories was promptly vetoed by Governor Chris Christie, and several gun dealers who have carried the Armatix alongside their other offerings have been targeted with boycotts and threats by hard-right gun activists.

The existence of smart gun technology in and of itself means nothing absent a significant investment of political capital to change producer, consumer, and regulatory practices. Ditto for other vaunted gun innovations. Mandating that guns include laser "microstamping" technology, for instance, could allow law enforcement to link any bullet casing found at a crime scene back to that weapon's owner—but only if civilian ownership were tracked and a database of gun IDs maintained. Meanwhile, in the real world, talk of gun owner registries is a political nonstarter, and the ATF's National Tracing Center, which would presumably maintain such a database, is so woefully underfunded that it still operates on microfilm and index cards.

Technology is no substitute for policy. And innovation is no substitute for a cultural sea change.

Behind all the political theater, money continues to flow. Politicians who condemn weapons manufacturers out of one side of their mouths lobby to fund firearms manufacturers in their districts out of the other. Even Chuck Schumer, who has loudly called for an assault weapons ban and vigorously condemned their manufacturers, has also pushed hard to award military contracts to domestic gun makers—and has even gone so far as to celebrate the job-creating "economic powerhouse" of assault rifle manufacturers in New York.

That these manufacturers have since followed tax incentives to more gun-friendly states is beside the point, as is the likelihood that Schumer supports arms manufacturers only insofar as their products wind up in the "right" hands of respectable civilians and police at home and American troops abroad. The bottom line is that beneath the rhetoric, high moral dudgeon, and misplaced hopes for utopian breakthroughs, the realities of political economy are what matter most.

Unfortunate pun aside, there is no magic bullet to "fix" this situation, one way or another. No brilliant inventor or charismatic entrepreneur can help us disrupt our way out of American capitalist militarism, legislative and regulatory capture, or interpersonal violence. The answers lie in the least flashy interventions: in old-fashioned harm reduction, in triage, in diminishing the inequalities that accelerate violence and precipitate lethal outcomes. Chasing dubious futures, we simply mortgage the present, and, in the most unthinking way, perpetuate it.

Civic Memory, Feminist Future

By Lidia Yuknavitch

(Originally appeared in
Electric Literature, April 2017)

1968

I am five and I'm the exact height of the top of the surgery scar on my mother's leg. I know this because she lets me sit on the toilet when she is getting dressed each morning. The scar is kid-high and the length of my torso. Pearly white railroad tracks covering her "birth defect." My mother is "disabled" though that word is never used in our house. My mother's leg is six inches shorter than the other leg. I don't know what "disabled" means but we heard it once when we did the March of Dimes walk door-to-door nice people giving donations something about children. I walk everywhere with my mother. We can't afford a car San Francisco is made of hills. Later in life I'll learn just how much walking all of those miles gave her pain in her leg.

Today we walk to a "polling station." We wait in the line. We walk up to a "polling booth." This year my mother begins to explain what "Democrats" are and she cries describing "assassinations." John F. Kennedy is a word I know. Shot dead. Dr. Martin Luther King Jr. is a word I know. Shot dead. My mother loved Bobby Kennedy the best. Shot dead. Nixon is a word I know my father swears at the television at night. God damn asshole is a word I know. The "Vietnam War" came into our living room through the television. Once I saw a fire hose and German Shepherds back some black people against a wall on the television and I cried and I couldn't understand why dogs would do that. Our dog named Maggie would never do that. My best friend Merrit is black and he wears a white shirt and tie almost every day to kindergarten. My teacher Mrs. Webb is black.

People make fun of me my hair too white people make fun of me I'm too shy people make fun of me I'm a crybaby. My favorite place to be is in a tree.

We go behind a plaid curtain plaid like my skirt. My mother pulls a lever. She whispers "Oh...Belle..." Belle is my nickname. My mother was born in Texas but she married a "Yankee" and left and never went back. Then she takes my hand and turns and opens the curtain.

I step with her into the future, a daughter and a mother moving though time and space, her lopsided heels clicking our misshapen path against the floor.

It will take years to undo that year. It will take lives.

1980

Seventeen is the heat and sweat of Florida and the rush of hormones but my desires move toward other girls about to be women and I do not have a prom queen body or a *Seventeen* body. I have an athlete body. Training two hours five thirty A.M. to seven thirty A.M. then high school then pool again four P.M. to six P.M. In between the waterworld hours I skip class with a boy gone to man who is infinitely more beautiful than I am. His brown skin his black hair his black eyes his perfect hands his desire for others like him. He makes art. He makes me a burgundy satin prom dress. My biceps bulge but I have no cleavage.

Seventeen is the cusp of everything. A girl's mind morphs toward woman faster than her brain can track, and so her body lurches and grinds forward more like an animal's. All around her images from her culture of what to be, what to look like, how to be wanted and thus counted.

At fifteen we moved from Seattle, Washington to Gainesville, Florida, my father told me that we moved so that I could train with the best swimming coach in the nation, Randy Reese. It was a big sacrifice they made. For me.

In reality, a place my father never lived, we moved because his engineering and architecture firm, CH2M Hill, transferred him. I'd already survived a childhood with my father. I'd survived an adolescence, though not without war wounds from the home front. My body carried a story underneath all the cover stories he told. From fifteen to seventeen I swam for my fucking life.

Why would a father tell a daughter that story? Why do men with power tell us stories away from what we know in our bodies? What do we do with the stories left ringing our ribs like tuning forks? Why do they lie?

At seventeen, I understand what voting is. I'm in honors English and

History classes. I know my turn is coming up. But that's not where my dreams live. All my kid life I dream of going to the Olympics.

At eighteen I'm on a high school relay team with the fastest time in the nation. Torry Blazey Holly Blair Michelle Reagan. 200 Yard Medley Relay. 1:47.620. All Americans listed in *Swimming World*.

The 1980 Summer Olympics boycott is one part of a number of actions initiated by President Jimmy Carter to protest the Soviet invasion of Afghanistan.

Where do the dreams of girls trying to swim or run from home go? Do they leave? Do our body stories make a nexus with the politics and histories of war and men? Or do they stay in our bodies, hidden like skin secrets, waiting for some future tense where we too might become realized?

What would a realized woman look like in America?

I don't vote.

I get a swimming scholarship from Texas Tech University, but in my heart I quit.

Ronald Reagan becomes president.[1]

1983

My first marriage dissolves.

My beautiful baby girl fish dies in my belly waters the day she is born. I leave reality and enter a real place called grief and psychosis. Everything is fractured water.

When I emerge, I am a writer; stories pouring from my body as if an entire ocean had been waiting there. Or maybe the voice of a girl—the one I was, the one who died in the belly of me, or the one who survived her father's hands and house—either way, she has an ungodly fire in her.

1984

I vote.

I vote because a new rage has emerged inside my body.

I have not yet learned that the rage is hope. I don't care about the democratic nominee for president, I care about Geraldine Ferraro. Geraldine Ferraro is the first woman presidential candidate representing a major American political party. I know that she is not the first woman to attempt ascension within the law and land and realm of the fathers, I know that Victoria Woodhull, whose running mate was Frederick Douglass, ran for president in 1872, for example, preceded her. She founded her own newspaper. She was the first woman to own a Wall Street investment firm. Douglass voted for Grant.

I know that Belva Ann Bennett Lockwood ran in 1884. She drafted the law passed by Congress which admitted women to practice before the Supreme Court; she then became the first woman lawyer to practice before the Court.

I know that Shirley Anita Chisholm was the first African American woman to seek a major party's nomination for U.S President in 1972. Before becoming the first black woman to serve in Congress, she was a school teacher and director of child care centers. I also know that Patsy Takemoto Mink ran in 1972, also one of the first women of color to serve in the U.S. Senate.

I know because I take women's studies classes in college, where I discover that this information has been left out of my entire American education. All those women's bodies. Voices. Stories.

The shape my rage takes is art. Drawing art painting art performance art and writing stories. So many stories are pouring out of my fingers I can't keep up with them. I don't even know where they are coming from, though I have a hunch they are coming from all the fathers who enact power on the bodies of others, biological fathers and state fathers and fatherlands and father heroes and father saviors and god the father from what's left of the Catholicism I was raised up and through. I am away from my father for the first time in my life. I suddenly wake up to the idea that America and democracy and capitalism are all about fathers. A certain idea of a father as head. Hero. Leader. Person with power. I denounce all fathers.

Rage blooms and grows in my mind and body and gut—where my daughter lived her entire life until her birthdeath—like an unapologetic violent flower. Thank oceans there is a word and action for it: feminism.

Ronald Reagan, formerly a Hollywood actor, is elected president. His face the word for it.

1989

The art I love most: Kathy Acker Karen Finley Lynne Tillman Laurie Anderson Andres Serrano Joel Peter Witkin Tim Miller Robert Mapplethorpe Holly Hughes Barbara Kruger Carolee Schneemann Cindy Sherman. In all of their work violence, death and sexuality kiss. Like in my life.

In 1989 two art pieces draw controversy to the NEA, Andres Serrano's *Piss Christ* and Robert Mapplethorpe's *The Perfect Moment*. Jesse Helmes emerges on the scene leading and effort to repress artistic production with obscenity laws.

In 1990 the NEA Four, performance artists Karen Finley, Tim Miller, John Fleck and Holly Hughes have their grants vetoed.

In March 1990 NEA grantees begin receiving a new clause in their agreements that states:

> *Public Law 101–121 requires that: None of the funds authorized to be appropriated for the National Endowment for the Arts ... may be used to promote, disseminate, or produce materials which in the judgement of the National Endowment for the Arts ... may be considered obscene, including but not limited to, depictions of sadomasochism, homoeroticism, the sexual exploitation of children, or individuals engaged in sex acts and which, when taken as a whole, do not have serious literary, artistic, political or scientific value.*

I hate Reagan. I hate "Morning in America," Ronald Reagan's reelection campaign which uses nostalgic images of America's heartland to help sell an optimistic future televised image.

I hate how the Reagan administration begins sending arms to Iran, via Isreal, in hopes that the weapons sales will lead the Iranians to pressure allies in Lebanon to release American hostages. The secret arms shipments violate Reagan's pledge never to negotiate with terrorists. Again.

Underneath that cover story Congress passes a law banning the diversion of US government funds to support Nicaragua's anticommunist Contra rebels. The Reagan administration violates the new law, leading to the Iran-Contra crisis. Reagan wins a second term in a historic landslide.

In 1986 a Lebanese magazine breaks the news that the U.S. has been secretly selling weapons to Iran. President Reagan delivers a nationally televised speech to address the Iran arms-for-hostages scandal. "Our government has a firm policy not to capitulate to terrorist demands," he says. "We did not—repeat, did not—trade weapons or anything else for hostages, nor will we."

Attorney General Edwin Meese, a staunch Reagan loyalist, begins an internal investigation into White House involvement in the Iran-Contra Scandal. Meese allows Iran-Contra conspirator Oliver North to shred thousands of potentially incriminating documents before they can be seized as evidence.

Meese finds administration officials guilty.

Marine Colonel Oliver North is fired.

In 1987 President Reagan goes on national TV to deliver a ridiculous apology for Iran-Contra: "A few months ago I told the American people I did not trade arms for hostages," he says. "My heart and my best intentions tell me that's true, but the facts and evidence tell me it's not."

The same year I watch Reagan give a speech in Berlin, telling Soviet Premier Gorbachev to "Tear down this wall." Where does he get off? People of color in my country are dying starving crawling from border to border.

Because I am high up in college, I'm reading about Marxism and Psycholinguistics and Semiotics and Feminisms—Eco Feminism and waves of Feminisms and Psychoanalytic Feminism and Marxist and Socialist Feminism and Cultural Feminism and Radical Feminism—Deconstruction and Literary Theory and Social Politics and Sociology. All I see is a sea of fathers and a runaway global kinesis made of money and power and guns. I understand this as capitalism. I understand my country as creating policies that have brutal, war-making and death-making consequences.

I understand my body as collateral damage.

My favorite writers are Ursula Le Guin and Doris Lessing and Toni Morrison and Leslie Marmon Silko and Elfriede Jelinek and Margaret Atwood and Maxine Hong Kingston and Marguerite Duras and Christa Wolf and Arundhati Roy. All of them telling and telling how the brutality of history is written on the bodies of the vulnerable and disenfranchised. How those bodies are walking skin maps of the spectacular and endless violences we commit culturally. Politically. Globally. All of them unwriting varieties of violence at the site of the body. All of them naming women and children as the necessary "matter" to colonize.

1989

I'm standing in a shitty voting booth in Eugene, Oregon. I vote so hard my eyes shiver. But even my heart sinks watching the televised image of Michael Dukakis riding around in an impotent tank. When Lloyd Bentsen debates what appears to be a moron-boy, Dan Quayle, and he delivers his best line, "Senator, I served with Jack Kennedy. I knew Jack Kennedy. Jack Kennedy was a friend of mine. Senator, you're no Jack Kennedy," I cry for nearly an hour.

Not because I could see the democrats were losing, but because memory came rushing back into my body. Loss. Grief. My mother. Her scar, her limp, the cancer that ate her lungs and breasts alive. The return of the repressed. A Nixon-ness coming back for revenge. The civil rights era being subsumed by consumer culture and capitalism. Television images forever asking us, are we dead yet? What brings us back to life and why? And when?

George H.W. Bush becomes president. But war was already being written across our bodies as well as the bodies of those we intended to colonize. Oil, power, land grabs, guns and money subsume all humanity, it seems.

I go to the courthouse in Eugene to protest Desert Storm every day and every night. No blood for oil. I'm accidentally on the cover of *Eugene Weekly*. I'm accidentally in a writing class with Ken Kesey. But I know there is no other father coming to save us.

1997

TESTIMONY OF PATRICK A. TRUEMAN

———

HON. JOHN T. DOOLITTLE of California, in the house of representatives

Tuesday, April 15, 1997

American Family Association:

> *Mr. Chairman and Members of the Committee: I want to thank you for the opportunity to appear before you today on behalf of American Family Association. As you are aware, for the past eight years AFA has been the leading organization opposing federal funding for the National Endowment for the Arts. In 1989, AFA president Rev. Donald Wildmon called to national attention the funding by the NEA of Andres Serrano's work "Piss Christ" which consisted of a crucifix submersed in the artists' urine. The fact that such a blasphemous work was federally funded outraged a great segment of American society and precipitated a battle to end federal funding of the agency. That battle will not end until funding for the NEA ends, rest assured of that fact....*

> *The threat that the NEA poses in the prosecution on obscenity and child pornography cases is not merely hypothetical. The difficulties I have outlined in this regard were faced by the U.S. Department of Justice during my years in the criminal division with respect to the funding by the NEA of an exhibit by the late Robert Mapplethorpe.*

The American Family Association is convinced after years of monitoring the NEA that the agency will never change. While it is only a small portion of its annual budget the NEA continues to fund pornographic works as "art." Some of the more recent and troubling works funded by the agency include grants to a group called FC2 and another called Women Make Movies, Inc. FC2 was provided $25,000 in the past year to support the publication of at least four books according to U.S. Representative Peter Hoekstra who has been tracking the NEA: S, by Jeffrey DeShell, Blood of Mugwump: A Tiresian Tale of Incest, by Doug Rice, Chick-Lit 2: No Chick Vics, edited by Cris Maza, Jeffrey Deshell and Elisabeth Sheffield and Mexico Trilogy, by D.N. Stuefloten. These books include descriptions of body mutilation, sadomasochistic sexual act, child sexual acts, sex between a nun and several priests, sodomy, incest, hetero and homosexual sex and numerous other graphically described sexual activities. Women Making Movies, Inc. received $112,700 in taxpayer money over the past three years for the production and distribution of several pornographic videos. Here are descriptions of but two taken from the groups catalog: "Ten Cents a Dance," a depiction of anonymous bathroom sex between two men; and another called "Sex Fish" which is "a furious montage of oral sex."

Oral sex is not art and the NEA and Congress should not pretend that it is. Please stop offending the taxpayers of America. Funding for the NEA should be eliminated.

I find my first literary tribe when I am published by Fiction Collective Two (FC2). Lance Olsen becomes a lifelong art and heart comrade. Ralph Berry becomes an experimental writing mentor to me. Jeffrey Deshell and Elisabeth Sheffield rise like human beacons illuminating for me how writing can work against the grain of cultural repression. Like Kathy Acker, Doug Rice blows the top of my head off by daring to place sexuality and the brutality of fathers and capitalism naked on a page; *Blood of Mugwump* becomes one of my favorite books. My stories also appear in Chick-Lit 2: No Chick Vics, edited by Cris Mazza, Jeffrey Deshell, and Elisabeth Sheffield.

Then congress holds hearings on the hill on the topic of art and obscenity, and all of our art flashes up like burning crosses.

2001

I vote for Bill Clinton twice. I know that he is a womanizer, I know that he lies, I know that the sentence "It depends upon what the meaning of the word 'is' is," is a crisis in representation. Everything we seem to have gained carries within it loss. Every birth has as its other, a death. My daughter's. Every sex act has my sexual history in it. My body carrying a culture.

Bill Clinton's presidency ends with a sex act.

George W. Bush is elected. A witless wealthy begetting of fathers.

My son Miles is born the year 9/11 happens. I'm breastfeeding Miles in our house in the woods. George W. Bush is on the television visiting a primary school in Florida when he gets the news. I see Miles' face at my breast. I see the face of Bush there on my TV, I see the faces of children. Seven minutes of Bush just sitting there processing what he's been told.

Suffer the children.

Who are we?

What is history? Does it live in our bodies or on our televisions? Do we move within it, or against it, or do we merely view it, consume it and move on? What can we move or not move? What part is ours to move?

2008

Did we choose the savior story? Did it save us? Has it ever? And what was the story underneath? And what work did we do to save ourselves and each other? Did we forget what might be coming?

2016

There is no photo for what my father did to his daughters.

It came into our bodies as a habit of being, a structure of conscious-ness, a way of life. Maybe it is akin to feeling discovered and conquered and colonized. Maybe the first colonizations are of the bodies of wom-en and children, and from there they extend like the outstretched hand of a man grabbing land. Cultures.

In my body my father and all the fathers after tried to annihilate my spirit.

He failed, but I carry the trace of the war in my skin song.

On the television a man stomps around behind a woman while she is trying to speak. I've seen this movie before. It's called my life. All the reasons people named for why Hillary Rodham Clinton was not good enough are exactly the same as everything ever said to women who were driven to stand in that place called power. Or was it simply self?

We did not survive all this way to let it ride.

2017

This is your present tense calling.

In some ways, I was born to do what's coming: to step exactly into the chaos and dark in order to live a life at all. There was no moment of my childhood that gave me joy except those rare pieces of time when I was out of the house of father. Thank oceans for swimming, for all the waters that have saved me.

We do know how to stand up. We do know how to hold a hand out for others, to make a human chain of hands and bodies against the wrong. We have to love the planet and animals and vulnerable people differently—with our whole bodies. No other now but this. The rest of life is reaching toward each other, loving into otherness, coming out of the dark, like some of us had to do as kids just to survive. If you forget, or if you become exhausted, ask your native brother or sister or people of color brother or sister or LGBT brother or sister. If this kid so scared she sometimes choked on her own breathing to get a word out of her mouth could do it, then all of us can.

If I could survive and escape my father's house, then it is possible for all of us.

[1] The year after my birth, Ronald Reagan appeared in his last film before he left Hollywood for politics. *The Killers*, based on a story by Earnest Hemingway, was a crime film starring Lee Marvin, John Cassavetes, Angie Dickinson, and Ronald Reagan. The movie remains notable for being Reagan's last theatrical film before entering politics, as well as the only one in which he plays a villain.

Science Fiction, Ancient Futures, and the Liberated Archive

By Walidah Imarisha

(This is an edited version of a keynote speech given at the Society of American Archivists Annual Conference, July 29, 2017, and transcribed by Irina Rogova.)

So often in social justice movements, we've been told to be realistic. What is a reasonable win? What little reform can we eke out of this system? And instead, I think it's important to recognize that all real, true, substantive social change was considered to be an impossibility before it happened. People were told it was science fiction, it was a fantasy. Eight-hour work day? Never gonna happen. Women getting the right to vote? Never gonna happen. The end of legal slavery in this country? Never gonna happen. Right? Folks said, "Well, we're not just gonna take your piecemeal reforms that you're offering, we're gonna dream that science fiction and make it reality." And they did.

But first you have to have that space to say, "What do we want?" Beyond the boundaries of what we're told is possible or what is real. We must create those sort of visionary spaces, to have conversations with each other within institutions and within communities, so that we can be developing the vision we're moving towards. It may exist beyond two-year grant cycles, or the funding cycle for universities. It may be a vision that's 20, 50, 100 years, but having those kind of shared visions give us something to collectively dream, and to collectively dream into being.

I've used archives and seen archives engaged in a lot of different ways, and have had amazing opportunities to see folks using archives in really liberated ways. I'm gonna just share some of those examples, to talk a little about those principles and values that I think are important around liberated archives. I'm gonna focus on three areas of my work. One is focusing around Oregon Black history—welcome to Oregon if you're not from here, it's about to get real. [Audience laughter.] Another area is the history of movements of resistance, movements for social justice, especially around challenging the prison industrial complex. And the last piece is about science fiction and organizing.

Around Oregon Black history—how many folks here are not from Oregon? Boom, alright. [Audience laughter.] Felt like we were doing the wave there for a minute. [Audience laughter.] I'll try to do the abbreviated version, and not the two-hour version where I'm like "And then y'all in 1965..." [Audience laughter.] Portland, especially, sort of exports this idea of Portlandia, this amazing liberal playground where it's the 1990s and you can... [Audience laughter] ...you can just have unicorns and birds on things. There is the image of Portland as the most progressive city in America, and that's definitely a part of how Portland markets and promotes itself. And then folks will say, "Well, there aren't a lot of people of color here, it's pretty white, but it's really progressive." I don't think those two things can coexist in the same space and time. If you are progressive, then at the core is justice. And if people of color cannot live here because of economics, because of laws, because of practices, because of feeling completely unwelcome, then you are not progressive.

The Oregon territory—which encompassed the entire Northwest and included Oregon, Washington, Idaho, parts of Wyoming and Montana—was founded as a racist white utopia. And so basically the call went out to white folks, "Come here and build the sort of white utopia you have dreamed of, away from the problems of the day." Which meant the presence of people of color. We know; we're good at reading code words. White folks flocked and answered that call. Because of this, many laws and practices were put into place, one of which was that Oregon was the only state in the Union admitted with a racial exclusionary clause in its constitution; a Black exclusion clause. So in Oregon's constitution, it said that Black folks were not allowed to live here, to hold property, or to make contracts. The very existence of being Black in Oregon was criminalized.

This was first passed before Oregon even became a state. The first Black exclusion law was passed in 1844, banning Blackness in essence in the entire Northwest. It also included the lash law that said that Black folks would be publicly whipped up to 39 lashes every 6 months until they left the Oregon territory. Then the other piece that's really important about this, is that this was in Oregon's constitution until 2001. So even though the law was repealed in 1926, the language was in the Constitution until 2001. Even though the Black community came to Oregon—through the legislature, through ballot boxes—again and again and again saying "Please remove this from your Constitution, and show us this is not the kind of place you want to live." And over and over again Oregon reinforced, "Actually this is exactly the kind of place we want to live." And it was a fight to get it removed in the 21st century; the efforts to remove the Black exclusion language did not pass easily.

But I think it's a very important principle and value, when we're talking about liberated archives, to make sure that we're always seeing oppressed people as active change-makers and not as passive victims. All of the advances in racial justice that have happened in this world have happened because of the courage, determination, and vision of people of color. Of the people who are most directly affected. People of color are active change-makers and leaders. They don't need saviors, they need allies. I think that that history is not something that is captured properly, and so you get things being put out—even documentaries—around some of this Black history that says the changes like the taking out of the language from the Oregon constitution was led by white legislators. They were like "Woo, we did it for you Black people! You're welcome!" [Audience laughter.] And it completely erases the fact that, again, Black folks for decades had been organizing under the most adverse and dangerous conditions for racial justice, including removing this language from the constitution.

Liberated archives have to center not only the voices and the experience but the leadership and vision of those people who are directly affected. It's not enough to just say, "Let's document the oppression that's happened." Let's view it through this empowerment lens so we can see the self-determination of oppressed people, and in this case of Black folks who made Oregon a better place for everyone. Again, the leadership of Black folks and people of color has made life better for all of us. So we have to make sure we are taking that leadership and highlighting that leadership at every level of creating liberatory archives or projects.

One of the collections that Portland State University has is the Charlotte Rutherford collection. Charlotte Rutherford was an amazing

Black leader. Her family has been organizing in the Black community in Oregon and Portland for 100 years. She worked as a community member and a leader, and she was determined to preserve the history of her work. Her family helped to found the Portland NAACP chapter, doing a lot of the work that was happening around civil rights. A lot of the papers are saran wrapped and pressed with an iron, a makeshift preservation technique to ensure those documents survived the test of time. That bootleg archivist. And I'm sure some of the archivists here are yelling inside, "No, you're ruining the documents!" But this shows the vision of Rutherford's family and the vision of these Black community leaders. This is at a time when the institutions that were archiving Oregon history wanted nothing to do with Oregon Black history. They didn't think it was important, didn't want those papers, didn't want those archives, didn't even return calls. Rutherford's family and others said, "This is important history. It's going to be necessary for the future. We're going to make sure it's saved and preserved."

This shows the importance of raising up the leadership, again, of folks who are marginalized, and especially people who sit at the intersections of identities and oppression. Charlotte Rutherford as a Black woman was holding this space, and so the vision we see is wider and broader because of that. It's incredibly important to make sure that those folks who are sitting at the intersections are at the table in the leadership thinking about what do we save and how.

To talk a little just about the history of social movements, I think that there are just a couple projects that I've used that I wanted to highlight that are based in communities. So there's the idea again of subverting existing archives but there are organizations like the Freedom Archives in San Francisco, which is an amazing resource that is archiving information about social movements, specifically around the 1960s-80s, really trying to archive the movement around Black and brown liberation, around women's rights, around queer rights, around sovereignty, around antiwar movements. And they have a physical space as well as a digital space. And they're producing new documentaries based on that to try to put this information out. They actually just finished a documentary about the Chicano Power movement. And this is really important because there is the significant lack of information around that, especially around the government repression. And so because they are a community based organization they heard from folks that this is what is needed, this is where there's a lack. They were able to respond to that and put it together in a way that is accessible to folks.

And that's the other piece, about making sure it's accessible to people. It's not just about saying, "Come, come into archives!" Even if it's open and it's free and you can just walk in, "You don't even need ID! Just come in, just put the gloves on, but you know it's cool you don't need ID!" [Audience laughter.] That's not enough because these institutions for centuries have been telling oppressed peoples, "You do not belong here." You can't change that just by sending out an e-mail and saying "Hey it's open!" and then sitting there and saying, "Why aren't people of color coming?" It's important to say: how do we make that this information is accessible, how do we take this knowledge that people *actually* want—not what we assume they want—out into the community where folks can use it and engage with it.

A lot of the work is around the prison industrial complex and challenging it. I think that there are amazing ways that folks are revisioning archives as living, breathing human beings as well, and I think that that's important. To recognize we are individually and collectively archives. To say, how do we capture that knowledge? How do we put it forward in an accessible way? And so just one example is the Story Telling and Organizing Project (STOP). They are collecting people's stories about how they dealt with harm or violence in their communities, in their families, outside of the criminal legal system. They did not want to call the police, they didn't wanna send folks to prison, but they wanted to address harm that was happening. They wanted to stop harm, or harm had come up, and they wanted to figure out what to do about that. And there's recordings and written transcripts of folks talking about this. I think that's incredibly important because also those folks are being shown as experts, so they're not just relaying what happened, they're relaying what happened and saying this is what we were thinking about it. Then they're analyzing it and showing the lessons they learned, the questions they still have about that, some things they learned since and ways they have implemented them and gotten different outcomes. So again, not only are folks active change-makers, they're also experts and should be respected as such. I think that that project is a really amazing one. It also really centers queer and trans people of color. Again, when we look at the intersections of oppression, we see the ways that folks have been resisting and building outside of the system.

There's an amazing project around alternatives to police by the Audre Lorde Project, which is queer and trans people of color, called Safe OUTside the System (SOS). I think that framework is so powerful, "safe outside the system." That for oppressed and marginalized people at the intersections, they've known there is no safety in this system. They know that calling the police actually often ends up with them arrested or assaulted. And so folks have been for decades, or for centuries, creating systems outside of these oppressive structures to keep one another safe. Looking at that is incredibly important because it's when we look at those intersections and the vision that folks have at those intersections that we see what true and total liberation can look like for all of us.

That idea that folks have been doing this for decades, for 100 years or more, that's where I want to bring us to and that's where the science fiction comes into play. I told you, wait for it, it's coming. But now I'll take a dramatic pause. [Audience laughter.]

So, as I said, I've done a lot of work around science fiction and organizing. I co-edited an anthology called *Octavia's Brood: Science Fiction Stories From Social Justice Movements*. It is...[Audience cheers]...oh, you guys...[Audience clapping, Walidah laughing.] It is science fiction and fantastical stories written by organizers, activists, and change-makers. Many of the contributors had never written fiction, let alone science fiction, but we knew they were holding these visionary spaces within the work that they were doing every single day. Because my co-editor, adrienne maree brown, and I strongly believe that all organizing is science fiction. Any time we imagine a world without war, a world without borders, without prisons, that is science fiction, because that world does not exist. But we can't build what we can't imagine. So we need visionary spaces, fantastical spaces, that as I said before allow us to throw out everything we're told is real and start with the question "What is the world that we want?" And then all we have to do is make it, you know? [Audience laughter.] Just like that, no biggie.

I think that a lot of times when we talk about science fiction, folks are thinking only about the future. And that's really rooted in a western colonial white supremacist ideology of history as this linear progression towards greatness. That the past is savagery and we're constantly getting better. Which is a justification for colonialism.

It's really important to recognize that linear time is a method of social control. That we're told this is all we have: "The past is gone, there's nothing it can do for you. The future is unknowable; you can't do anything about that. All you have is the present. So just be as happy as you can, get as much as you can, because that's all you have." And instead, many societies historically, specifically brown cultures, have recognized that time is not linear. It's circular, it's spherical. It exists in multiple places at once. We live in the past, the present, and the future simultaneously. And this is quantum physics now. We're all like, "Welcome to the game, quantum physicists, we been here! But thanks for now acknowledging and validating what we've known forever."

This has a very strong impact on this idea of liberated archives, because we need that past to move forward into the future. We need these visions of liberation that existed before. We need to be able to study them; we need to be able to explore them. We need to be able to say, "What is the wisdom and knowledge that exists in the past that will help us build the future?" These are a lot of principles of Afrofuturism or Black futurism. The ancient future. We're not saying that the past is savagery and the future is perfection where we live in chrome and glass floating sky cities. But instead saying our ancestors' knowledge and visions and dreams are part of the expertise needed to advance. Our ancestors existed in the past, the present, and the future at once, so recognizing how they saw the world, how they envisioned the future, and how they lived the future, is something that can help us to create the future that we want.

Archives are key to that, because it's not just about having the information—it's about having those tangible objects, those pieces that you can hold and touch and in many ways time travel through. I was teaching at Stanford this last year, and they have Black Panther leader Huey Newton's archives. And I was holding Huey Newton's diary. I'm reading his amazing political exploration of apartheid. And then he has some pressed flowers on the next page. Pressed flowers, y'all! [Audience laughter.] Y'all know, everyone knows, the image of Huey Newton in the rattan chair with the gun and the spear, which is amazing. Then it's juxtaposed with those beautiful little pressed flowers in his diary.

This is why archives are important, because we're whole human beings. And if we only see these flat images, we can't build a future out of that. We can't build a future just out of a gun and a spear and a rattan chair, right? I'm not saying we don't need the gun and the spear and the rattan chair—but we need the pressed flowers too. We need to be able to see ourselves as whole, and we need to be able to envision a future where we can be whole.

I want to share an excerpt from one of the stories from *Octavia's Brood*, by Alexis Pauline Gumbs. Alexis is an amazing Black feminist, queer visionary, academic, and storyteller, and she's doing her own kind of Black feminist storytelling archives project. The story is called "Evidence," and it's really about time traveling through research, through archives, as academic practice. And so in this story, it's someone in the future researching our time, in a future where liberation has happened. They got free in the future, so the character is writing to the past to say, "By the way, everything's awesome!" It all works out. Y'all, no spoilers, but it all works out. The character is researching this time, but she's also sending messages back to this time, and Alexis's question with the piece is can we, through sort of archives, academic research, scholarship, actually reach back into the past and change it? Not just learn from the past, but actually impact it to ensure the future that we want. Not just theoretically, but in real tangible ways.

This is the beginning to "Evidence" by Alexis Pauline Gumbs:

> *By reading past this point, you agree that you are accountable to the council. You affirm our collective agreement that in the time*

of accountability, the time past law and order, the story is the storehouse of justice. You remember that justice is no longer punishment. You affirm that the time of crime was an era of refused understanding and stunted evolution. We believe now in the experience of brilliance on the scale of the intergalactic tribe.

Today the evidence we need is legacy. May the public record show and celebrate that Alandrix consciously exists in an ancestral context. May this living textual copy of her digital compilation and all its future amendments be a resource for Alandrix, her mentors, her loved ones and partners, her descendants, and detractors to use in the ongoing process of supporting her just intentions.

We are grateful that you are read this. Thank you for remembering.

With love and what our ancestors called 'faith,'

the intergenerational council of possible elders.

I want that future so badly, y'all. This story reinforces for us that we must be asking what are our principles and values? We don't have to figure out completely what that future looks like, but a key step to doing that is asking how do we want to be. How do we want to be, individually and collectively? If we solidify those principles, then we will always be moving in the right direction.

We, all of us, are a collective archive. We have countless and endless quantum possibilities for liberation, but it does take the incorporation of those values and principles of self-determination, of autonomy, of empowerment, of decolonization, and of liberated vision. The past is one of our countless starting points. Liberated archives can hold that space, hold that ancient future. That ancient future that can be our every day lived reality, if we are ready to do the beautiful, hard, glorious, empowering work of building the world we dream of into existence.

Thank y'all so much, I hope you have an amazing day. [Audience clapping/standing ovation.]

On Liking Women

By Andrea Long Chu

(Originally appeared in
N+1, Issue 30, Winter 2018)

O nce a week, for a single semester of high school, I would be dismissed early from class to board the athletics bus with fifteen teenage girls in sleek cap-sleeved volleyball jerseys and short shorts. I was the only boy.

Occasionally a girl who still needed to change would excuse herself behind a row of seats to slip out of her school uniform into the team's dark-blue colors. For more minor wardrobe adjustments, I was simply asked to close my eyes. In theory, all sights were trained on the game ahead where I, as official scorekeeper, would push numbers around a byzantine spreadsheet while the girls leapt, dug, and dove with raw, adolescent power. But whatever discipline had instilled itself before a match would dissolve in its aftermath, often following a pit stop for greasy highway-exit food, as the girls relaxed into an innocent dishabille: untucked jerseys, tight undershirts, the strap of a sports bra. They talked, with the candor of postgame exhaustion, of boys, sex, and other vices; of good taste and bad blood and small, sharp desires. I sat, and I listened, and I waited, patiently, for that wayward electric pulse that passes unplanned from one bare upper arm to another on an otherwise unremarkable Tuesday evening, the away-game bus cruising back over the border between one red state and another.

The truth is, I have never been able to differentiate liking women from wanting to be like them. For years, the former desire held the latter in its mouth, like a capsule too dangerous to swallow. When I trawl the seafloor of my childhood for sunken tokens of things to come, these bus rides are about the gayest thing I can find. They probably weren't even all that gay. It is common, after all, for high school athletes to try to squash the inherent homoeroticism of same-sex sport under the heavy cleat of denial. But I'm too desperate to salvage a single genuine lesbian memory from the wreckage of the scared, straight boy whose life I will never not have lived to be choosy. The only other memory with a shot at that title is my pubescent infatuation with my best friend, a moody, low-voiced, Hot Topic–shopping girl who, it dawned on me only many years later, was doing her best impression of Shane from *The L Word*. One day she told me she had a secret to tell me after school; I spent the whole day queasy with hope that a declaration of her affections was forthcoming. Later, over the phone, after a pause big enough to drown in, she told me she was gay. "I thought you might say that," I replied, weeping inside. A decade later, after long having fallen out of touch, I texted her. "A week ago, I figured out that I am trans," I wrote. "You came out to me all those years ago. Just returning the favor."

This was months before I began teaching my first undergraduate recitation, where for the second time in my life—but the first time as a woman—I read Valerie Solanas's *SCUM Manifesto*. The *SCUM Manifesto* is a deliciously vicious feminist screed calling for the revolutionary overthrow of all men; Solanas self-published it in 1967, one year before she shot Andy Warhol on the sixth floor of the Decker Building in New York City. I wondered how my students would feel about it. In the bathroom before class, as I fixed my lipstick and fiddled with my hair, I was approached by a thoughtful, earnest young woman who sat directly to my right during class. "I loved the Solanas reading," she told me breathlessly. "I didn't know that was a thing you could study." I cocked my head, confused. "You didn't know what was a thing you could study?" "Feminism!" she said, beaming. In class, I would glance over at this student's notes, only to discover that she had filled the page with the word *SCUM*, written over and over with the baroque tenderness usually reserved for the name of a crush.

I, too, had become infatuated with feminism in college. I, too, had felt the thrill of its clandestine discovery. I had caught a shy glimpse of her across a dim, crowded dormitory room vibrating with electronic music and unclear intentions: a low-key, confident girl, slightly aloof, with a gravity all neighboring bodies obeyed. Feminism was too cool, too effortlessly hip, to be interested in a person like me, whom social anxiety had prevented from speaking over the telephone until well into high school. Besides, I heard she only dated women. I limited myself, therefore, to acts of distant admiration. I left critical comments on the student newspaper's latest exposé of this or that frat party. I took a Women's Studies course that had only one other man in it. I read desperately, from Shulamith Firestone to Jezebel, and I wrote: bizarre, profane plays about rape culture, one where the archangel Gabriel had a monologue so vile it would have burned David Mamet's tongue clean off; and ugly, strange poetry featuring something I was calling the Beautiful Hermaphrodite Proletariat. Feminism was all I wanted to think about, talk about. When I visited home, my mother and my sister, plainly irritated, informed me that I did not know what it was like to be a woman. But a crush was a crush, if anything buttressed by the conviction that feminism, like any of the girls I had ever liked, was too good for me.

It was in my junior year of college that I first read the *SCUM Manifesto*, crossing over the East River in a lonely subway car. It exhilarated me: the grandeur, the brutal polemics, the raw, succulent style of the whole thing. Solanas was *cool*. Rereading *SCUM*, I realized this was no accident. The manifesto begins like this:

> Life in this society being, at best, an utter bore and no aspect of society being at all relevant to women, there remains to civic-minded, responsible, thrill-seeking females only to overthrow the government, eliminate the money system, institute complete automation and destroy the male sex.

What's striking here is not Solanas's revolutionary extremism per se, but the flippancy with which she justifies it. Life under male supremacy isn't oppressive, exploitative, or unjust: it's just fucking boring. For Solanas, an aspiring playwright, politics begins with an aesthetic judgment. This is because male and female are essentially styles for her, rival aesthetic schools distinguishable by their respective adjectival palettes. Men are timid, guilty, dependent, mindless, passive, animalistic, insecure, cowardly, envious, vain, frivolous, and weak. Women are strong, dynamic, decisive, assertive, cerebral, independent, self-confident, nasty, violent, selfish, freewheeling, thrill-seeking, and arrogant. Above all, women are cool and groovy.

"...what few men remain
after the revolution will be
generously permitted to wither
away on drugs or in drag,
grazing in pastures or hooked into
twenty-four-hour feeds..."

Yet as I read back through the manifesto in preparation for class, I was surprised to be reminded that, for all her storied manhating, Solanas is surprisingly accommodating in her pursuit of male extinction. For one thing, the groovy, freewheeling females of Solanas's revolutionary infantry SCUM (which at one point stood for "Society for Cutting Up Men," though this phrase appears nowhere within the manifesto) will spare any man who opts to join its Men's Auxiliary, where he will declare himself "a turd, a lowly abject turd." For another, what few men remain after the revolution will be generously permitted to wither away on drugs or in drag, grazing in pastures or hooked into twenty-four-hour feeds allowing them to vicariously live the high-octane lives of females in action. And then there's this:

> If men were wise, they would seek to become really female,
> would do intensive biological research that would lead to
> men, by means of operations on the brain and nervous
> system, being able to be transformed in psyche, as well as
> body, into women.

This line took my breath away. This was a vision of transsexuality as separatism, an image of how male-to-female gender transition might express not just disidentification with maleness but disaffiliation with men. Here, transition, like revolution, was recast in aesthetic terms, as if transsexual women decided to transition, not to "confirm" some kind of innate gender identity, but because being a man is stupid and boring.

I overread, perhaps. In 2013, an event in San Francisco intended as a tribute to Solanas on the twenty-fifth anniversary of her death was canceled after bitter conflict broke out on its Facebook page over what some considered Solanas's transphobia. One trans woman described having been harassed in queer spaces by radical feminists who referenced Solanas almost as often as they did Janice Raymond, whose 1979 book *The Transsexual Empire: The Making of the She-Male* is a classic of anti-trans feminism. Others went on the offensive. Mira

Bellwether, creator of *Fucking Trans Women*, the punk-rock zine that taught the world to muff, wrote a lengthy blog post explaining her misgivings about the event, characterizing the *SCUM Manifesto* as "potentially the worst and most vitriolic example of lesbian-feminist hate speech" in history. She goes on to charge Solanas with biological essentialism of the first degree, citing the latter's apparent appeal to genetic science: "The male is a biological accident: the Y (male) gene is an incomplete X (female) gene, that is, it has an incomplete set of chromosomes. In other words, the male is an incomplete female, a walking abortion, aborted at the gene stage." For Bellwether, this is unequivocal proof that everything *SCUM* says about men, it also says about trans women.

Yet these are odd accusations. To call Solanas a "lesbian feminist" is to imply, erroneously, that she was associated with lesbian groups like New York City's Lavender Menace, which briefly hijacked the Second Congress to Unite Women in 1970 to protest homophobia in the women's movement and distribute their classic pamphlet "The Woman-Identified Woman." But Solanas was neither a political lesbian nor a lesbian politico. She was by all accounts a loner and a misfit, a struggling writer and sex worker who sometimes identified as gay but always looked out for number one. The dedication to her riotous 1965 play *Up Your Ass* reads, "I dedicate this play to ME, a continuous source of strength and guidance, and without whose unflinching loyalty, devotion, and faith this play would never have been written." (It was this play, whose full title is *Up Your Ass, or From the Cradle to the Boat, or The Big Suck, or Up from the Slime*, that Solanas tried first to sweet-talk, then to strong-arm, Andy Warhol into producing.)

As for the matter of genetics, I suppose I ought to be offended to have my Y chromosomes' good name raked through the mud. Frankly, though, I have a hard time getting it up for a possession I consider about as valuable as a $15 gift card to Blockbuster. The truth is, if it's hard for contemporary readers to tell men and trans women apart in Solanas's analysis, it is not because she thinks all trans women are men; if anything, it's because she thinks all men are closeted trans women. When Solanas hisses that maleness is a "deficiency disease," I am reminded of those trans women who diagnose themselves, only half-jokingly, with testosterone poisoning. When she snarls that men are "biological accidents," all I hear is the eminently sensible claim that every man is *literally* a woman trapped in the wrong body. This is what the *SCUM Manifesto* calls pussy envy, from which all men suffer, though few dare to admit it aside from "faggots" and "drag queens" whom Solanas counts among the least miserable of the lot. Hence the sentiment Solanas expresses through Miss Collins, one of two quick-witted queens who grace the filthy pages of *Up Your Ass*:

> MISS COLLINS: Shall I tell you a secret? I despise men. Oh, why do I have to be one of them? (*Brightening.*) Do you know what I'd like more than anything in the world to be? A Lesbian. Then I could be the cake and eat it too.

Bellwether might object that I am, again, being too generous. But generosity is the only spirit in which a text as hot to the touch as the *SCUM Manifesto* could have ever been received. This is after all a pamphlet advocating mass murder, and what's worse, property damage. It's not as if those who expressed their disappointment over the tribute's cancellation did so in blanket approval of Solanas's long-term plans for total human extinction (women included) or her attempted murder of a man who painted soup cans. As Breanne Fahs recounts in her recent biography of Solanas, the shooting was the straw that broke the back of the camel known as the National Organization for Women (NOW), which despite its infancy—it was founded in 1966, only two years earlier—had already suffered fractures over abortion and lesbianism. As the radical feminists Ti-Grace Atkinson and Florynce Kennedy visited Solanas in prison, the latter agreeing to represent Valerie pro bono, then president Betty Friedan scrambled to distance NOW from what she viewed as a problem that most certainly had a name, demanding in a telegram that Kennedy "DESIST IMMEDIATELY FROM LINKING NOW IN ANY WAY WITH VALERIE SOLANAS." Within the year, both Kennedy and Atkinson had left the organization, each going on to found their own, ostensibly more radical groups: the Feminist Party and the October 17th Movement, respectively. Likewise, after the Solanas tribute was canceled in 2013, folks hoping to hash out the Facebook fracas in person held a splinter event called "We Who Have Complicated Feelings About Valerie Solanas."

This is simply to note that disagreement over Solanas's legacy is an old feminist standard, the artifact of a broader intellectual habit that critiques like Bellwether's lean on. This is the thing we call feminist historiography, with all its waves and groups and fabled conferences. Any good feminist bears stitched into the burning bra she calls her heart that tapestry of qualifiers we use to tell one another stories about ourselves and our history: radical, liberal, neoliberal, socialist, Marxist, separatist, cultural, corporate, lesbian, queer, trans, eco, intersectional, anti-porn, anti-work, pro-sex, first-, second-, third-, sometimes fourth-wave. These stories have perhaps less to do with What Really Happened than they do with what Fredric Jameson once called "the 'emotion' of great historiographic form"—that is, the satisfaction of synthesizing the messy empirical data of the past into an elegant historical arc in which everything that happened could not have happened otherwise.

To say, then, that these stories are rarely if ever "true" is not merely to repeat the axiom that taxonomy is taxidermy, though it cannot be denied that the objects of intellectual inquiry are forever escaping, like B-movie zombies, from the vaults of their interment. It is also to say that all cultural things, *SCUM Manifesto* included, are answering machines for history's messages at best only secondarily. They are rather, first and foremost, occasions for people to feel something: to adjust the pitch of a desire or up a fantasy's thread count, to make overtures to a new way to feel or renew their vows with an old one. We read things, watch things, from political history to pop culture, as feminists and as people, because we want to belong to a community or public, or because we are stressed out at work, or because we are looking for a friend or a lover, or perhaps because we are struggling to figure out how to feel political in an age and culture defined by a general shipwrecking of the beautiful old stories of history.

So when Bellwether condemns the *SCUM Manifesto* as "the pinnacle of misguided and hateful 2nd wave feminism and lesbian-feminism," this condemnation is a vehicle for a kind of political disappointment that feminists are fond of cultivating with respect to preceding generations of feminists. In this version of the story, feminism excluded trans women in the past, is learning to include trans women now, and will center trans women in the future. This story's plausibility is no doubt due to a dicey bit of revisionism implied by the moniker *trans-exclusionary radical feminist*, often shortened to TERF. Like most kinds of feminist, TERFs are not a party or a unified front. Their beliefs, while varied, mostly boil down to a rejection of the idea that transgender women are, in fact, women. They also don't much like the name TERF, which they take to be a slur—a grievance that would be beneath contempt if it weren't also true, in the sense that all bywords for bigots are intended to be defamatory. The actual problem with an epithet like TERF is its historiographic sleight of hand: namely, the erroneous implication that all TERFs are holdouts who missed the third wave, old-school radical feminists who never learned any better. This permits their being read as a kind of living anachronism through which the past can be discerned, much as European anthropologists imagined so-called primitive societies to be an earlier stage of civilizational development caught in amber.

In fact, we would do better to talk about TERFs in the context of the internet, where a rebel alliance of bloggers like Feminist Current's Meghan Murphy and GenderTrender's Linda Shanko spend their days shooting dinky clickbait at the transsexual empire's thermal exhaust ports. The true battles rage on Tumblr, in the form of comments, memes, and doxing; it is possible, for instance, to find Tumblrs entirely devoted to cataloging other Tumblr users who are known "gender critical feminists," as they like to refer to themselves. But this conflict has as much to do with the ins and outs of social media—especially Tumblr, Twitter, and Reddit—as it does with any great ideological conflict. When a subculture espouses extremist politics, especially online, it is tempting but often incorrect to take those politics for that subculture's beating heart. It's worth considering whether TERFs, like certain strains of the alt-right, might be defined less by their political ideology (however noxious) and more by a complex, frankly fascinating relationship to trolling, on which it will be for future anthropologists, having solved the problem of digital ethnography, to elaborate.

Of course, feminist transphobia is no more an exclusively digital phenomenon than white nationalism. There were second-wave feminists who sincerely feared and hated trans women. Some of them are even famous, like the Australian feminist Germaine Greer, author of the 1974 best seller *The Female Eunuch*. Few TERFs curl their lips with Greer's panache. This is how she described an encounter with a fan, in the *Independent* magazine in 1989:

> On the day that *The Female Eunuch* was issued in America, a person in flapping draperies rushed up to me and grabbed my hand. "Thank you," it breathed hoarsely, "Thank you so much for all you've done for us girls!" I smirked and nodded and stepped backward, trying to extricate my hand from the enormous, knuckly, hairy, be-ringed paw that clutched it. The face staring into mine was thickly coated with pancake make-up through which the stubble was already burgeoning, in futile competition with a Dynel wig of immense luxuriance and two pairs of false eyelashes. Against the bony ribs that could be counted through its flimsy scarf dress swung a polished steel women's liberation emblem. I should have said, "You're a man. *The Female Eunuch* has done less than nothing for you. Piss off."

Little analysis is needed to show that disgust like Greer's belongs to the same traffic in woman-hating she and her fellow TERFs supposedly abhor. Let us pause instead to appreciate how rarely one finds transmisogyny, whose preferred medium is the spittle of strangers, enjoying the cushy stylistic privileges of middlebrow literary form. It's like watching Julia Child cook a baby.

Then again, Greer has long imagined herself as feminism's id, periodically digging herself out of the earth to rub her wings together and molt on network television. In 2015, she made waves when she criticized as "misogynist" *Glamour* magazine's decision to give their Woman of the Year award to Caitlyn Jenner, then fresh off her *Vanity Fair* photo shoot. In response to the backlash, Greer released this gem of a statement: "Just because you lop off your dick and then wear a dress doesn't make you a fucking woman. I've asked my doctor to give me long ears and liver spots and I'm going to wear a brown coat but that won't turn me into a fucking cocker spaniel." More surprising is when a second-wave icon like Atkinson, onetime defender of Solanas, trots out TERF talking points at a Boston University conference in 2014: "There is a conflict around gender. That is, feminists are trying to get rid of gender. And transgendered [sic] reinforce gender." That Atkinson's remarks arrived at a conference whose theme was "Women's Liberation in the Late 1960s and Early 1970s" only encourages wholesale dismissals of the second wave as the Dark Ages of feminist history.

Yet consider the infamous West Coast Lesbian Conference of 1973. The first night of the conference, the transsexual folk singer Beth Elliott's scheduled performance was interrupted by protesters who tried to kick her off the stage. The following day, the radical feminist Robin Morgan, editor of the widely influential 1970 anthology *Sisterhood Is Powerful*, delivered a hastily rewritten keynote in which she unloaded on Elliott, calling her "an opportunist, an infiltrator, and a destroyer—with the mentality of a rapist." Morgan's remarks were soon printed in the short-lived underground newspaper *Lesbian Tide*, where they could enjoy a wider audience:

> I will not call a male "she"; thirty-two years of suffering in this androcentric society, and of surviving, have earned me the title "woman"; one walk down the street by a male transvestite, five minutes of his being hassled (which he may enjoy), and then he dares, he dares to think he understands our pain? No, in our mothers' names and in our own, we must not call him sister. We know what's at work when whites wear blackface; the same thing is at work when men wear drag.

This is where reports of the conference usually end, often with a kind of practiced sobriety about How Bad Shit Was. Yet as the historian Finn Enke argues in an excellent article forthcoming in *Transgender Studies Quarterly*, many accounts leave out the fact that the San Fran-

cisco chapter of the national lesbian organization Daughters of Bili-
tis had welcomed a 19-year-old Beth Elliott in 1971 after her parents
rejected her, that Elliott had been elected chapter vice president that
same year, that she had been embraced by the Orange County Dyke
Patrol at the Gay Women's Conference in Los Angeles, and that she
had been *a member of the organizing committee* for the very conference
where her presence was disputed by a vocal minority of attendees. As
for the vitriolic keynote, Enke suggests that Morgan's attacks on Elliott were born of the former's insecurity over being invited to speak
at a conference for lesbians despite her being shacked up with a man,
whose effeminacy she often tried, unsuccessfully, to parlay into a basis for her own radical credentials.

This is to say two things. First, the radical feminism of the Sixties
and Seventies was as mixed a bag as any political movement, from
Occupy to the Bernie Sanders campaign. Second, at least in this case,
feminist transphobia was not so much an expression of anti-trans an-
imus as it was an indirect, even peripheral repercussion of a much
larger crisis in the women's liberation movement over how people
should go about feeling political. In expanding the scope of feminist
critique to the terrain of everyday life—a move which produced a
characteristically muscular brand of theory that rivaled any Marxist's
notes on capitalism—the second wave had inadvertently painted itself into a corner. If, as radical feminist theories claimed, patriarchy
had infested not just legal, cultural, and economic spheres but the
psychic lives of *women themselves*, then feminist revolution could only
be achieved by combing constantly through the fibrils of one's consciousness for every last trace of male supremacy—a kind of political
nitpicking, as it were. And nowhere was this more urgent, or more difficult, than the bedroom. Fighting tirelessly for the notion that sex was
fair game for political critique, radical feminists were now faced with
the prospect of putting their mouths where their money had been.
Hence Atkinson's famous slogan: "Feminism is the theory, lesbianism
is the practice." This was the political climate in which *both* Elliott
and Morgan, as a transsexual woman and a suspected heterosexual
woman, respectively, could find their statuses as legitimate subjects of
feminist politics threatened by the incipient enshrining, among some

radical feminists, of something called lesbianism as the preferred aesthetic form for mediating between individual subjects and the history they were supposed to be making—call these the personal and the political.

So while radical feminism as a whole saw its fair share of trans-loving lesbians and trans-hating heterosexuals alike, there *is* a historical line to be traced from political lesbianism, as a specific, by no means dominant tendency *within* radical feminism, to the contemporary phenomenon we've taken to calling trans-exclusionary radical feminism. Take Sheila Jeffreys, an English lesbian feminist recently retired from a professorship at the University of Melbourne in Australia. In her salad days, Jeffreys was a member of the Leeds Revolutionary Feminist Group, remembered for its fiery conference paper "Political Lesbianism: The Case Against Heterosexuality," published in 1979. The paper defined a political lesbian as "a woman-identified woman who does not fuck men" but stopped short of mandating homosexual sex. The paper also shared the *SCUM Manifesto*'s dead-serious sense of humor: "Being a heterosexual feminist is like being in the resistance in Nazi-occupied Europe where in the daytime you blow up a bridge, in the evening you rush to repair it." These days, Jeffreys has made a business of abominating trans women, earning herself top billing on the TERF speaking circuit. Like many TERFs, she believes that trans women's cheap imitations of femininity (as she imagines them) reproduce the same harmful stereotypes through which women are subordinated in the first place. "Transgenderism on the part of men," Jeffreys writes in her 2014 book *Gender Hurts*, "can be seen as a ruthless appropriation of women's experience and existence." She is also fond of citing sexological literature that classifies transgenderism as a paraphilia. It is a favorite claim among TERFs like Jeffreys that transgender women are gropey interlopers, sick voyeurs conspiring to infiltrate women-only spaces and conduct the greatest panty raid in military history.

I happily consent to this description. Had I ever been so fortunate as to attend the legendarily clothing-optional Michigan Womyn's Music Festival before its demise at the hands of trans activists in 2015, you can bet your Birkenstocks it wouldn't have been for the music.

"We are separatists from our own bodies. We are militants of so fine a caliber that we regularly take steps to poison the world's supply of male biology."

Indeed, at least among lesbians, trans-exclusionary radical feminism might best be understood as gay panic, girl-on-girl edition. The point here is not that all TERFs are secretly attracted to trans women—though so delicious an irony undoubtedly happens more often than anyone would like to admit—but rather that trans-exclusionary feminism has inherited political lesbianism's dread of desire's ungovernability. The traditional subject of gay panic, be he a US senator or just a member of the House, is a subject menaced by his own politically compromising desires: to preserve himself, he projects these desires onto another, whom he may now legislate or gay-bash out of existence. The political lesbian, too, is a subject stuck between the rock of politics and desire's hard place. As Jeffreys put it in 2015, speaking to the Lesbian History Group in London, political lesbianism was intended as a solution to the all-too-real cognitive dissonance produced by heterosexual feminism: "Why go to all these meetings where you're creating all this wonderful theory and politics, and then you go home to, in my case, Dave, and you're sitting there, you know, in front of the telly, and thinking, 'It's *weird*. This feels *weird*.'" But true separatism doesn't stop at leaving your husband. It proceeds, with paranoid rigor, to purge the apartments of the mind of anything remotely connected to patriarchy. Desire is no exception. Political lesbianism is founded on the belief that even desire becomes pliable at high enough temperatures. For Jeffreys and her comrades, lesbianism was not an innate identity, but an act of political will. This was a world in which biology was not destiny, a world where being a lesbian was about what got you woke, not wet.

Only heterosexuality might not have been doing it for Dave, either. It seems never to have occurred to Jeffreys that some of us "transgenders," as she likes to call us, might opt to transition precisely in order to escape from the penitentiary she takes heterosexuality to be. It is a supreme irony of feminist history that there is no woman more woman-identified than a gay trans girl like me, and that Beth Elliott and her sisters were the OG political lesbians: women who had walked away from both the men in their lives and the men whose lives they'd been living. We are separatists from our own bodies. We are militants of so fine a caliber that we regularly take steps to poison the world's supply of male biology. To TERFs like Jeffreys, we say merely that imitation is the highest form of flattery. But let's keep things in perspective. Because of Jeffreys, a few women in the Seventies got haircuts. Because of us, there are literally *fewer men on the planet*. Valerie, at least, would be proud. The Society for Cutting Up Men is a rather fabulous name for a transsexual book club.

But now I really am overreading. That trans lesbians should be pedestaled as some kind of feminist vanguard is a notion as untenable as it is attractive. In defending it, I would be neglecting what I take to be the true lesson of political lesbianism as a failed project: that nothing good comes of forcing desire to conform to political principle. You could sooner give a cat a bath. This does not mean that politics has no part to play in desire. Solidarity, for instance, can be terribly arousing—this was no doubt one of the best things the consciousness-raising groups of the Seventies had going for them. But you can't get aroused *as an act of* solidarity, the way you might stuff envelopes or march in the streets with your sisters-in-arms. Desire is, by nature, childlike and chary of government. The day we begin to qualify it by the righteousness of its political content is the day we begin to prescribe some desires and prohibit others. That way lies moralism only. Just try to imagine life as a feminist anemone, the tendrils of your desire withdrawing in an instant from patriarchy's every touch. There would be nothing to watch on TV.

It must be underscored how unpopular it is on the left today to countenance the notion that transition expresses not the truth of an identity but the force of a desire. This would require understanding transness as a matter not of who one *is*, but of what one *wants*. The primary function of gender identity as a political concept—and, increasingly, a legal one—is to bracket, if not to totally deny, the role of desire in the thing we call gender. Historically, this results from a wish among transgender advocates to quell fears that trans people, and trans women in particular, go through transition in order to *get stuff*: money, sex, legal privileges, little girls in public restrooms. As the political theorist Paisley Currah observes in his forthcoming book, the state has been far more willing to recognize sex reclassification when the reclassified individuals don't get anything out of it. In 2002, the Kansas Supreme Court voided the marriage of a transsexual woman and her then-deceased cisgender husband, whose $2.5 million estate she was poised to inherit, on the grounds that their union was invalid under Kansas's prohibition on same-sex marriage. The sex on the woman's Wisconsin birth certificate, which she had successfully changed from M to F years earlier, now proved worthless when she tried to cash it in.

Now I'm not saying I think that this woman transitioned to get rich quick. What I am saying is, *So what if she had?* I doubt that any of us transition simply because we want to "be" women, in some abstract, academic way. I certainly didn't. I transitioned for gossip and compliments, lipstick and mascara, for crying at the movies, for being someone's girlfriend, for letting her pay the check or carry my bags, for the benevolent chauvinism of bank tellers and cable guys, for the telephonic intimacy of long-distance female friendship, for fixing my makeup in the bathroom flanked like Christ by a sinner on each side, for sex toys, for feeling hot, for getting hit on by butches, for that secret knowledge of which dykes to watch out for, for Daisy Dukes, bikini tops, and all the dresses, and, my god, *for the breasts*. But now you begin to see the problem with desire: we rarely want the things we should. Any TERF will tell you that most of these items are just the traditional trappings of patriarchal femininity. She won't be wrong, either. Let's be clear: TERFs are gender abolitionists, even if that abolitionism is a shell corporation for garden-variety moral disgust. When it comes to the question of feminist revolution, TERFs leave trans girls like me in the dust, primping. In this respect, someone like Ti-Grace Atkinson, a self-described radical feminist committed to the revolutionary dismantling of gender as a system of oppression, is not the dinosaur; I, who get my eyebrows threaded every two weeks, am.

Perhaps my consciousness needs raising. I muster a shrug. When the airline loses your luggage, you are not making a principled political statement about the tyranny of private property; you just want your goddamn luggage back. This is most painfully evident in the case of bottom surgery, which continues to baffle a clique of queer theorists who, on the strength and happenstance of a shared prefix, have been all too ready to take transgender people as mascots for their politics of transgression. These days, the belief that getting a vagina will make you into a real woman is retrograde in the extreme. Many good feminists still only manage to understand bottom surgery by qualifying it as a personal aesthetic choice: *If that's what makes you feel more comfortable in your body, that's great.* This is as wrongheaded as it is condescending. To be sure, gender confirmation surgeries are aesthetic practices, continuous with rather than distinct from the so-called cosmetic surgeries. (No one goes into the operating room asking for an ugly cooch.) So it's not that these aren't aesthetic decisions; it's that they're not *personal*. That's the basic paradox of aesthetic judgments: they are, simultaneously, subjective and universal. Transsexual women don't want bottom surgery because their personal opinion is that a vagina would look or feel better than a penis. Transsexual women want bottom surgery because *most women have vaginas.* Call that transphobic if you like—that's not going to keep me from Chili's-Awesome-Blossoming my dick.

I am being tendentious, dear reader, because I am trying to tell you something that few of us dare to talk about, especially in public, especially when we are trying to feel political: not the fact, boringly obvious to those of us living it, that many trans women wish they were cis women, but the darker, more difficult fact that many trans women *wish they were women, period*. This is most emphatically not something trans women are supposed to want. The grammar of contemporary trans activism does not brook the subjunctive. Trans women *are* women, we are chided with silky condescension, as if we have all confused ourselves with Chimamanda Ngozi Adichie, as if we were all simply trapped in the wrong politics, as if the cure for dysphoria were wokeness. How can you want to be something you already are? Desire implies deficiency; want implies want. To admit that what makes women like me transsexual is not identity but desire is to admit just how much of transition takes place in the waiting rooms of wanting things, to admit that your breasts may never come in, your voice may never pass, your parents may never call back.

Call this the romance of disappointment. You want something. You have found an object that will give you what you want. This object is a person, or a politics, or an art form, or a blouse that fits. You attach yourself to this object, follow it around, carry it with you, watch it on TV. One day, you tell yourself, it will give you what you want. Then, one day, it doesn't. Now it dawns on you that your object will probably never give you what you want. But this is not what's disappointing, not really. What's disappointing is what happens next: nothing. You keep your object. You continue to follow it around, stash it in a drawer, water it, tweet at it. It still doesn't give you what you want—but you knew that. You have had another realization: not getting what you want has very little to do with wanting it. Knowing better usually doesn't make it better. You don't want something because wanting it will lead to getting it. You want it because you want it. This is the zero-order disappointment that structures all desire and makes it possible. After all, if you could only want things you were guaranteed to get, you would never be able to want anything at all.

This is not to garner pity for sad trannies like me. We have enough roses by our beds. It is rather to say, minimally, that trans women want things too. The deposits of our desire run as deep and fine as any. The richness of our want is staggering. Perhaps this is why coming out can feel like crushing, why a first dress can feel like a first kiss, why dysphoria can feel like heartbreak. The other name for disappointment, after all, is love.

The Future in Motion: Why I Judge High School Debate Tournaments

By Bryan Washington

(Originally published on
Catapult.com, May 2017)

Every now and again, whenever I'm back in Texas, I'll judge Student Congress tournaments on Houston's north side, which is the side without the money and the dazzle and the glitz; and even though the lockers are fading, and the water fountains are busted, black and brown kids from the houses lining the feeder road dress up in teal-blue button-downs and Dockers and cufflinks to debate abortion rights and immigration directives and incarceration reduction and statewide food deserts, turning their underfunded, overcrowded, often neglected schools into the intellectual boons our country could have if it actually wanted them.

In the 115th United States Congress, there are currently fifteen Asian Americans. We've got thirty-eight Latinos, forty-nine African Americans, two Muslims, 104 women, thirty Jewish lawmakers, and seven LGBT representatives. Individually, the numbers look remarkable. Laudable, even. And of the 535 members between the House and the Senate, they are still hardly indicative of their constituency's diversity.

But the debates provide a look at a parallel reality. On any given weekend in Alief, you'll end up judging a room full of Latino kids, all of them fluent in the "One China" policy. Or a roster that's half-Nigerian, riffing on public school funding. Or twenty Vietnamese kids poking at cyber security. Or a few trans students. Or a gaggle of atheists. Or the kids who simply don't know who they'd like to be just yet, only that they are Something Else, and they'll let you know when they figure it out.

It is science fiction, in a literal sense. It's also fucking absurd. And it's as clear a portrait of our nation as any you're likely to find.

One time this black girl collapsed in tears over a hypothetical amendment to the Paris Agreement. But she wasn't particularly upset. She

" They were, as far as I was concerned, the future in motion."

was just making her point. Clean air is delicious, she said, inhaling for effect. She slapped her hands after every syllable.

One time I watched a kid stand on a chair to recite the Pledge of Allegiance backwards. She was blonde, and pale, and wholly convinced that our current naturalization process was dehumanizing.

One time I heard this Native American kid give a speech in defense of Black History Month. He stood up in his suit jacket, navy blue with polka dots all over. He couldn't have been taller than five-foot-three. He'd nearly buzzed his hair to the scalp. He wore a snapback and outrageously tight khakis and I actually couldn't imagine what his life outside that classroom was like. But he cleared his throat, and rose both palms, citing Zora Neale Hurston and Mae Carol Jemison and Baldwin and Obama and Left Eye. The argument was immaculate. His competitors held their breath. The woman judging beside me circled his name until it bled through her scorecard.

After the round had come to a close, I stopped the kid at the door. I told him what he'd done was electrifying. He'd given me a light for the dark.

He looked at me like I was bugging. Thanks, he said, but it's just common sense.

If you judge enough tournaments, you start running into the same kids. Some of them are theatrical, outsized in the halls. Some of them are known in the orbit of their particular debate circuit. It's not rare to catch one kid trying on a particular identity, or a way of speaking or presenting themselves, only to abandon it a few weekends later for some whole other thing.

Some kids, you just watch them, and you know they're too big for their situation. You hope and hope they'll get their chance and you know it's possible they won't.

On paper, the winners are the kids who make the most compelling arguments. They're judged by a group of older folks and their peers. But really, it's all about how a kid works the room. It's less about what they're saying than how they say it, and who they're saying it to.

Some kids have the most compelling arguments, the most solid evidence, and they'll end up missing the point entirely. They'll lose the round for being an asshole in their delivery. And I've seen rounds where the kid's argument isn't all that sharp, with evidence that's dubious at best, but they're up-front with their audience, and straight about where they're coming from, and their peers will see the human in their argument and all of a sudden they'll end up nabbing a trophy.

Sometimes a tournament with lower-income schools will be held in a wealthy district. It's just how the schedule works out. The kids' buses roll in, graying and dented, and the lots that they'll park in are blow-dried and lacquered. You'll catch the contempt on some of their faces. Some are a little awestruck. Others find it fucking hilarious.

For a lot of these students, it's their first contact with Money, and people who've known nothing outside of it. The first time they've met kids whose schools support independent budgets for debate, with boosters and fundraisers and rallies. Those kids will have the tailored suits and the swagger and everything else, but their appearance isn't necessarily a predictor of success. More often than not, it isn't any indicator at all.

Once, after an especially compelling argument in favor of expanding the military's budget, delivered in a tone you'd be hard-pressed to call anything but entitled, the blonde kid who made it sat right down when a black girl in cornrows rose her hand to make a comment. She wore a purple dress. Her sandals slapped at the tile.

Your argument, she said, makes no fucking sense. None.

And then she sat down. Because that was it. And the blonde kid looked embarrassed, just like the judges looked embarrassed, and the girl looked embarrassed for anyone who thought that it did.

If a student asks why I'm judging—and they almost never do—I tell them that I used to run with STUCO myself. It's not like I get paid (if anything, I lose money). I hit my first debate tournament for the same reason I did everything else at fourteen: Some guy I was into was doing it. I didn't say shit my entire first round. I just sat there, staring. Lost in the sauce.

I'd entered the room with people that I'd thought were just like me, but who, in actuality, were operating in another milieu entirely. They were learned and wise. Wholly inscrutable. I couldn't tell you which countries fought in World War I, let alone negotiate our ongoing conflicts in the Middle East.

At the end of the set, one of the judges found me. He was this black guy, tall and stocky. He said he wished I'd spoken. He wanted to hear what I had to say. But maybe next time, he said, smiling at the thought, and the memory of that conversation shamed me for weeks.

The year after that, I actually spoke once or twice. I advanced in a few tournaments. The year after that, I won district. The year after that, I was an East Texas representative, and the National Tournament was held in Dallas, and I came in contact with Old Money for the first time in my life.

Every other kid wore an American flag pin on their lapels, and I'd driven into the city in a hoodie and sweats. Once I'd been knocked out in the first round, a complete and immediate annihilation, and the girl who'd sealed my fate called me in the commons afterwards. She was tall and dark-skinned. She said she hadn't meant to do it. She hadn't known I was the only other one.

I smiled and shook her hand.

I'd already driven back to Houston and a few months had passed before I realized what she'd meant.

One time there was a qualifying tournament, and I'd been asked to stand in as a judge. It was held in a very white part of town, and the room we were in mirrored the majority. There were two black kids on the roster, but they weren't doing all that great. It was clear they weren't going to make it. Their involvement was very low-stakes.

Still, once a resolution on the Civil Rights Act entered the docket, they both became animated; they flipped the whole fucking script.

Their arguments were emphatic. They were physically emoting. They looked drained once they'd finished, like this was the thing they'd been waiting all year to do.

This is why we're here, said one of the kids, beating at her chest. This is why we're here, we're here to make a change.

It was a hypothetical argument in a well-manicured classroom in a part of the city whose social justice engagement was flailing at best; but, for a minute, it felt like the world. For a minute, it felt like these kids held our future in their hands.

It's been a minute since I've judged a tournament, but I keep track of some of the students. They'll add me on Facebook or Twitter. Some of them have gone off to university. They'll DM me for recommendation letters, or references, or advice, and the most I've got to tell them is that I don't know what I'm doing either.

One got in touch with me a little while ago. He figured we'd grab lunch the last time I was in Texas. He'd taken a gap year to float around Europe, and then he prolonged his studies for another year, and now he was considering bypassing university entirely. He wanted to talk it out with someone. I put that meeting off for weeks, and then one day, months after he'd messaged, I said I'd meet him in town.

When I made it to the diner, he was sitting with this other kid. They had notes all over the table. They gestured wildly, breathlessly. The other kid looked like he really believed; they looked like they were getting real work done, and watching the pair from around the corner I seriously considered leaving them be.

There wasn't a damn thing I could give them that they weren't already giving themselves. They were, as far as I was concerned, the future in motion. They were the best of us, of where we are headed. But eventually, they looked up, and I waved from across the room, and I wandered across the restaurant to see what they could teach me.

Someone to Watch Over Me: What Happens When We Let Tech Care for Our Aging Parents

By Lauren Smiley

(Originally appeared in WIRED Magazine, January 2018)

A rlyn Anderson grasped her father's hand and presented him with the choice. "A nursing home would be safer, Dad," she told him, relaying the doctors' advice. "It's risky to live here alone—"

"No way," Jim interjected. He frowned at his daughter, his brow furrowed under a lop of white hair. At 91, he wanted to remain in the woodsy Minnesota cottage he and his wife had built on the shore of Lake Minnetonka, where she had died in his arms just a year before. His pontoon—which he insisted he could still navigate just fine—bobbed out front.

Arlyn had moved from California back to Minnesota two decades earlier to be near her aging parents. Now, in 2013, she was fiftysomething, working as a personal coach, and finding that her father's decline was all-consuming.

Her father—an inventor, pilot, sailor, and general Mr. Fix-It; "a genius," Arlyn says—started experiencing bouts of paranoia in his mid-eighties, a sign of Alzheimer's. The disease had progressed, often causing his thoughts to vanish mid-sentence. But Jim would rather risk living alone than be cloistered in an institution, he told Arlyn and her older sister, Layney. A nursing home certainly wasn't what Arlyn wanted for him either. But the daily churn of diapers and cleanups, the carousel of in-home aides, and the compounding financial strain (she had already taken out a reverse mortgage on Jim's cottage to pay the caretakers) forced her to consider the possibility.

Jim, slouched in his recliner, was determined to stay at home. "No way," he repeated to his daughter, defiant. Her eyes welled up and she hugged him. "OK, Dad." Arlyn's house was a 40-minute drive from the cottage, and for months she had been relying on a patchwork of technology to keep tabs on her dad. She set an open laptop on the counter so she could chat with him on Skype. She installed two cameras, one in his kitchen and another in his bedroom, so she could check whether the caregiver had arrived, or God forbid, if her dad had fallen. So when she read in the newspaper about a new digital eldercare service called CareCoach a few weeks after broaching the subject of the nursing home, it piqued her interest. For about $200 a month, a human-powered avatar would be available to watch over a homebound person 24 hours a day; Arlyn paid that same amount for just nine hours of in-home help. She signed up immediately.

A Google Nexus tablet arrived in the mail a week later. When Arlyn plugged it in, an animated German shepherd appeared on-screen, standing at attention on a digitized lawn. The brown dog looked cutesy and cartoonish, with a bubblegum-pink tongue and round, blue eyes.

She and Layney visited their dad later that week, tablet in hand. Following the instructions, Arlyn uploaded dozens of pictures to the service's online portal: images of family members, Jim's boat, and some of his inventions, like a computer terminal known as the Teleray and a seismic surveillance system used to detect footsteps during the Vietnam War. The setup complete, Arlyn clutched the tablet, summoning the nerve to introduce her dad to the dog. Her initial instinct that the service could be the perfect companion for a former technologist had splintered into needling doubts. Was she tricking him? Infantilizing him?

Tired of her sister's waffling, Layney finally snatched the tablet and presented it to their dad, who was sitting in his armchair. "Here, Dad, we got you this." The dog blinked its saucer eyes and then, in Google's female text-to-speech voice, started to talk. Before Alzheimer's had taken hold, Jim would have wanted to know exactly how the service worked. But in recent months he'd come to believe that TV characters were interacting with him: A show's villain had shot a gun at him, he said; Katie Couric was his friend. When faced with an onscreen character that actually was talking to him, Jim readily chatted back.

Jim named his dog Pony. Arlyn perched the tablet upright on a table in Jim's living room, where he could see it from the couch or his recliner. Within a week Jim and Pony had settled into a routine, exchanging pleasantries several times a day. Every 15 minutes or so

Pony would wake up and look for Jim, calling his name if he was out of view. Sometimes Jim would "pet" the sleeping dog onscreen with his finger to rustle her awake. His touch would send an instantaneous alert to the human caretaker behind the avatar, prompting the Care-Coach worker to launch the tablet's audio and video stream. "How are you, Jim?" Pony would chirp. The dog reminded him which of his daughters or in-person caretakers would be visiting that day to do the tasks that an onscreen dog couldn't: prepare meals, change Jim's sheets, drive him to a senior center. "We'll wait together," Pony would say. Often she'd read poetry aloud, discuss the news, or watch TV with him. "You look handsome, Jim!" Pony remarked after watching him shave with his electric razor. "You look pretty," he replied. Sometimes Pony would hold up a photo of Jim's daughters or his inventions between her paws, prompting him to talk about his past. The dog complimented Jim's red sweater and cheered him on when he struggled to buckle his watch in the morning. He reciprocated by petting the screen with his index finger, sending hearts floating up from the dog's head. "I love you, Jim!" Pony told him a month after they first met—something CareCoach operators often tell the people they are monitoring. Jim turned to Arlyn and gloated, "She does! She thinks I'm real good!"

About 1,500 miles south of Lake Minnetonka, in Monterrey, Mexico, Rodrigo Rochin opens his laptop in his home office and logs in to the CareCoach dashboard to make his rounds. He talks baseball with a New Jersey man watching the Yankees; chats with a woman in South Carolina who calls him Peanut (she places a cookie in front of her tablet for him to "eat"); and greets Jim, one of his regulars, who sips coffee while looking out over a lake.

Rodrigo is 35 years old, the son of a surgeon. He's a fan of the Spurs and the Cowboys, a former international business student, and a bit of an introvert, happy to retreat into his sparsely decorated home office each morning. He grew up crossing the border to attend school in McAllen, Texas, honing the English that he now uses to chat with elderly people in the United States. Rodrigo found CareCoach on an

"Having Pony there eased her anxiety about leaving [her father] alone, and the virtual dog's small talk lightened the mood."

online freelancing platform and was hired in December 2012 as one of the company's earliest contractors, role-playing 36 hours a week as one of the service's avatars.

In person, Rodrigo is soft-spoken, with wire spectacles and a beard. He lives with his wife and two basset hounds, Bob and Cleo, in Nuevo León's capital city. But the people on the other side of the screen don't know that. They don't know his name—or, in the case of those like Jim who have dementia, that he even exists. It's his job to be invisible. If Rodrigo's clients ask where he's from, he might say MIT (the CareCoach software was created by two graduates of the school), but if anyone asks where their pet actually is, he replies in character: "Here with you."

Rodrigo is one of a dozen CareCoach employees in Latin America and the Philippines. The contractors check on the service's seniors through the tablet's camera a few times an hour. (When they do, the dog or cat avatar they embody appears to wake up.) To talk, they type into the dashboard and their words are voiced robotically through the tablet, designed to give their charges the impression that they're chatting with a friendly pet. Like all the CareCoach workers, Rodrigo keeps meticulous notes on the people he watches over so he can coordinate their care with other workers and deepen his relationship with them over time—this person likes to listen to Adele, this one prefers Elvis, this woman likes to hear Bible verses while she cooks. In one client's file, he wrote a note explaining that the correct response to "See you later, alligator" is "After a while, crocodile." These logs are all available to the customer's social workers or adult children, wherever they may live. Arlyn started checking Pony's log between visits with her dad several times a week. "Jim says I'm a really nice person," reads one early entry made during the Minnesota winter. "I told Jim that he was my best friend. I am so happy."

After watching her dad interact with Pony, Arlyn's reservations about outsourcing her father's companionship vanished. Having Pony there eased her anxiety about leaving Jim alone, and the virtual dog's small talk lightened the mood.

Pony was not only assisting Jim's human caretakers but also inadvertently keeping an eye on them. Months before, in broken sentences, Jim had complained to Arlyn that his in-home aide had called him a bastard. Arlyn, desperate for help and unsure of her father's recollection, gave her a second chance. Three weeks after arriving in the house, Pony woke up to see the same caretaker, impatient. "Come on, Jim!" the aide yelled. "Hurry up!" Alarmed, Pony asked why she was screaming and checked to see if Jim was OK. The pet—actually, Rodrigo—later reported the aide's behavior to CareCoach's CEO, Victor Wang, who emailed Arlyn about the incident. (The caretaker knew there was a human watching her through the tablet, Arlyn says, but may not have known the extent of the person's contact with Jim's family behind the scenes.) Arlyn fired the short-tempered aide and started searching for a replacement. Pony watched as she and Jim conducted the interviews and approved of the person Arlyn hired. "I got to meet her," the pet wrote. "She seems really nice."

Pony—friend and guard dog—would stay.

Victor Wang grew up feeding his Tamagotchis and coding choose-your-own-adventure games in QBasic on the family PC. His parents moved from Taiwan to suburban Vancouver, British Columbia, when Wang was a year old, and his grandmother, whom he called Lao Lao in Mandarin, would frequently call from Taiwan. After her husband died, Lao Lao would often tell Wang's mom that she was lonely, pleading with her daughter to come to Taiwan to live with her. As she grew older, she threatened suicide. When Wang was 11, his mother moved back home for two years to care for her. He thinks of that time as the honey-sandwich years, the food his overwhelmed father packed him each day for lunch. Wang missed his mother, he says, but adds, "I was never raised to be particularly expressive of my emotions."

At 17, Wang left home to study mechanical engineering at the University of British Columbia. He joined the Canadian Army Reserve, serving as an engineer on a maintenance platoon while working on his undergraduate degree. But he scrapped his military future when, at 22, he was admitted to MIT's master's program in mechanical engineering. Wang wrote his dissertation on human-machine interaction, studying a robotic arm maneuvered by astronauts on the International Space Station. He was particularly intrigued by the prospect of harnessing tech to perform tasks from a distance: At an MIT entrepreneurship competition, he pitched the idea of training workers in India to remotely operate the buffers that sweep US factory floors.

In 2011, when he was 24, his grandmother was diagnosed with Lewy body dementia, a disease that affects the areas of the brain associated with memory and movement. On Skype calls from his MIT apartment, Wang watched as his grandmother grew increasingly debilitated. After one call, a thought struck him: If he could tap remote labor to sweep far-off floors, why not use it to comfort Lao Lao and others like her?

Wang started researching the looming caretaker shortage in the US—between 2010 and 2030, the population of those older than 80 is projected to rise 79 percent, but the number of family caregivers available is expected to increase just 1 percent.

In 2012 Wang recruited his cofounder, a fellow MIT student working on her computer science doctorate named Shuo Deng, to build CareCoach's technology. They agreed that AI speech technology was too rudimentary for an avatar capable of spontaneous conversation tailored to subtle mood and behavioral cues. For that, they would need humans.

Older people like Jim often don't speak clearly or linearly, and those with dementia can't be expected to troubleshoot a machine that misunderstands. "When you match someone not fully coherent with a device that's not fully coherent, it's a recipe for disaster," Wang says. Pony, on the other hand, was an expert at deciphering Jim's needs. Once, Pony noticed that Jim was holding onto furniture for support, as if he were dizzy. The pet persuaded him to sit down, then called Arlyn. Deng figures it'll take about 20 years for AI to be able to master that kind of personal interaction and recognition. That said, the Care-Coach system is already deploying some automated abilities. Five years ago, when Jim was introduced to Pony, the offshore workers behind the camera had to type every response; today CareCoach's software creates roughly one out of every five sentences the pet speaks. Wang aims to standardize care by having the software manage more of the patients' regular reminders—prodding them to take their medicine, urging them to eat well and stay hydrated. CareCoach workers are part freewheeling raconteurs, part human natural-language processors, listening to and deciphering their charges' speech patterns or nudging the person back on track if they veer off topic. The company recently began recording conversations to better train its software in senior speech recognition.

CareCoach found its first customer in December 2012, and in 2014 Wang moved from Massachusetts to Silicon Valley, renting a tiny office space on a lusterless stretch of Millbrae near the San Francisco airport. Four employees congregate in one room with a view of the parking lot, while Wang and his wife, Brittany, a program manager he met at a gerontology conference, work in the foyer. Eight tablets with sleeping pets onscreen are lined up for testing before being shipped to their respective seniors. The avatars inhale and exhale, lending an eerie sense of life to their digital kennel.

Wang spends much of his time on the road, touting his product's health benefits at medical conferences and in hospital executive suites. Onstage at a gerontology summit in San Francisco last summer, he deftly impersonated the strained, raspy voice of an elderly man talking to a CareCoach pet while Brittany stealthily cued the replies from her laptop in the audience. The company's tablets are used by hospitals and health plans across Massachusetts, California, New York, South Carolina, Florida, and Washington state. Between corporate and individual customers, CareCoach's avatars have interacted with hundreds of users in the US. "The goal," Wang says, "is not to have a little family business that just breaks even."

The fastest growth would come through hospital units and health plans specializing in high-need and elderly patients, and he makes the argument that his avatars cut health care costs. (A private room in a nursing home can run more than $7,500 a month.) Preliminary research has been promising, though limited. In a study conducted by Pace University at a Manhattan housing project and a Queens hospital, CareCoach's avatars were found to reduce subjects' loneliness, delirium, and falls. A health provider in Massachusetts was able to replace a man's 11 weekly in-home nurse visits with a CareCoach tablet, which diligently reminded him to take his medications. (The man told nurses that the pet's nagging reminded him of having his wife back in the house. "It's kind of like a complaint, but he loves it at the same time," the project's lead says.) Still, the feelings aren't always so cordial: In the Pace University study, some aggravated seniors with dementia lashed out and hit the tablet. In response, the onscreen pet sheds tears and tries to calm the person.

More troubling, perhaps, were the people who grew too fiercely attached to their digital pets. At the conclusion of a University of Washington CareCoach pilot study, one woman became so distraught at the thought of parting with her avatar that she signed up for the service, paying the fee herself. (The company gave her a reduced rate.) A user in Massachusetts told her caretakers she'd cancel an upcoming vacation to Maine unless her digital cat could come along.

We're still in the infancy of understanding the complexities of aging humans' relationship with technology. Sherry Turkle, a professor of social studies, science, and technology at MIT and a frequent critic of tech that replaces human communication, described interactions between elderly people and robotic babies, dogs, and seals in her 2011 book, *Alone Together*. She came to view roboticized eldercare as a cop-out, one that would ultimately degrade human connection. "This kind of app—in all of its slickness and all its 'what could possibly be wrong with it?' mentality—is making us forget what we really know about what makes older people feel sustained," she says: caring, interpersonal relationships. The question is whether an attentive avatar makes a comparable substitute. Turkle sees it as a last resort. "The assumption is that it's always cheaper and easier to build an app than to have a conversation," she says. "We allow technologists to propose the unthinkable and convince us the unthinkable is actually the inevitable."

But for many families, providing long-term in-person care is simply unsustainable. The average family caregiver has a job outside the home and spends about 20 hours a week caring for a parent, according to AARP. Nearly two-thirds of such caregivers are women. Among eldercare experts, there's a resignation that the demographics of an aging America will make technological solutions unavoidable. The number of those older than 65 with a disability is projected to rise from 11 million to 18 million from 2010 to 2030. Given the option, having a digital companion may be preferable to being alone. Early research shows that lonely and vulnerable elders like Jim seem content to communicate with robots. Joseph Coughlin, director of MIT's AgeLab, is pragmatic. "I would always prefer the human touch over a robot," he says. "But if there's no human available, I would take high tech in lieu of high touch."

CareCoach is a disorienting amalgam of both. The service conveys the perceptiveness and emotional intelligence of the humans powering it but masquerades as an animated app. If a person is incapable of consenting to CareCoach's monitoring, then someone must do so on their behalf. But the more disconcerting issue is how cognizant these seniors are of being watched over by strangers. Wang considers his product "a trade-off between utility and privacy." His workers are trained to duck out during baths and clothing changes.

Some CareCoach users insist on greater control. A woman in Washington state, for example, put a piece of tape over her CareCoach tablet's camera to dictate when she could be viewed. Other customers like Jim, who are suffering from Alzheimer's or other diseases, might not realize they are being watched. Once, when he was temporarily placed in a rehabilitation clinic after a fall, a nurse tending to him asked Arlyn what made the avatar work. "You mean there's someone overseas looking at us?" she yelped, within earshot of Jim. (Arlyn isn't sure whether her dad remembered the incident later.) By default, the app explains to patients that someone is surveilling them when it's first introduced. But the family members of personal users, like Arlyn, can make their own call.

Arlyn quickly stopped worrying about whether she was deceiving her dad. Telling Jim about the human on the other side of the screen "would have blown the whole charm of it," she says. Her mother had Alzheimer's as well, and Arlyn had learned how to navigate the disease: Make her mom feel safe; don't confuse her with details she'd have trouble understanding. The same went for her dad. "Once they stop asking," Arlyn says, "I don't think they need to know anymore." At the time, Youa Vang, one of Jim's regular in-person caretakers, didn't comprehend the truth about Pony either. "I thought it was like

Siri," she said when told later that it was a human in Mexico who had watched Jim and typed in the words Pony spoke. She chuckled. "If I knew someone was there, I may have been a little more creeped out."

Even CareCoach users like Arlyn who are completely aware of the person on the other end of the dashboard tend to experience the avatar as something between human, pet, and machine—what some roboticists call a third ontological category. The caretakers seem to blur that line too: One day Pony told Jim that she dreamed she could turn into a real health aide, almost like Pinocchio wishing to be a real boy.

Most of CareCoach's 12 contractors reside in the Philippines, Venezuela, or Mexico. To undercut the cost of in-person help, Wang posts English-language ads on freelancing job sites where foreign workers advertise rates as low as $2 an hour. Though he won't disclose his workers' hourly wages, Wang claims the company bases its salaries on factors such as what a registered nurse would make in the CareCoach employee's home country, their language proficiencies, and the cost of their internet connection.

The growing network includes people like Jill Paragas, a CareCoach worker who lives in a subdivision on Luzon island in the Philippines. Paragas is 35 years old and a college graduate. She earns about the same being an avatar as she did in her former call center job, where she consoled Americans irate about credit card charges. ("They wanted to, like, burn the company down or kill me," she says with a mirthful laugh.) She works nights to coincide with the US daytime, typing messages to seniors while her 6-year-old son sleeps nearby.

Before hiring her, Wang interviewed Paragas via video, then vetted her with an international criminal background check. He gives all applicants a personality test for certain traits: openness, conscientiousness, extroversion, agreeableness, and neuroticism. As part of the CareCoach training program, Paragas earned certifications in delirium and dementia care from the Alzheimer's Association, trained in US health care ethics and privacy, and learned strategies for counseling those with addictions. All this, Wang says, "so we don't get anyone who's, like, crazy." CareCoach hires only about 1 percent of its applicants.

Paragas understands that this is a complicated business. She's befuddled by the absence of family members around her aging clients. "In my culture, we really love to take care of our parents," she says. "That's why I'm like, 'She is already old, why is she alone?'" Paragas has no doubt that, for some people, she's their most significant daily relationship. Some of her charges tell her that they couldn't live without her. Even when Jim grew stubborn or paranoid with his daughters, he always viewed Pony as a friend. Arlyn quickly realized that she had gained a valuable ally.

As time went on, the father, daughter, and family pet grew closer. When the snow finally melted, Arlyn carried the tablet to the picnic table on the patio so they could eat lunch overlooking the lake. Even as Jim's speech became increasingly stunted, Pony could coax him to talk about his past, recounting fishing trips or how he built the house to face the sun so it would be warmer in winter. When Arlyn took her dad around the lake in her sailboat, Jim brought Pony along. ("I saw mostly sky," Rodrigo recalls.)

One day, while Jim and Arlyn were sitting on the cottage's paisley couch, Pony held up a photograph of Jim's wife, Dorothy, between her paws. It had been more than a year since his wife's death, and Jim hardly mentioned her anymore; he struggled to form coherent sentences. That day, though, he gazed at the photo fondly. "I still love her," he declared. Arlyn rubbed his shoulder, clasping her hand over her mouth to stifle tears. "I am getting emotional too," Pony said. Then Jim leaned toward the picture of his deceased wife and petted her face with his finger, the same way he would to awaken a sleeping Pony.

When Arlyn first signed up for the service, she hadn't anticipated that she would end up loving—yes, loving, she says, in the sincerest sense of the word—the avatar as well. She taught Pony to say "Yeah, sure, you betcha" and "don't-cha know" like a Minnesotan, which made her laugh even more than her dad. When Arlyn collapsed onto the couch after a long day of caretaking, Pony piped up from her perch on the table:

"Arnie, how are you?"

Alone, Arlyn petted the screen—the way Pony nuzzled her finger was weirdly therapeutic—and told the pet how hard it was to watch her dad lose his identity.

"I'm here for you," Pony said. "I love you, Arnie."

When she recalls her own attachment to the dog, Arlyn insists her connection wouldn't have developed if Pony was simply high-functioning AI. "You could feel Pony's heart," she says. But she preferred to think of Pony as her father did—a friendly pet—rather than a person on the other end of a webcam. "Even though that person probably had a relationship to me," she says, "I had a relationship with the avatar."

Still, she sometimes wonders about the person on the other side of the screen. She sits up straight and rests her hand over her heart. "This is completely vulnerable, but my thought is: Did Pony really care about me and my dad?" She tears up, then laughs ruefully at herself, knowing how weird it all sounds. "Did this really happen? Was it really a relationship, or were they just playing solitaire and typing cute things?" She sighs. "But it seemed like they cared."

When Jim turned 92 that August, as friends belted out "Happy Birthday" around the dinner table, Pony spoke the lyrics along with them. Jim blew out the single candle on his cake. "I wish you good health, Jim," Pony said, "and many more birthdays to come."

In Monterrey, Mexico, when Rodrigo talks about his unusual job, his friends ask if he's ever lost a client. His reply: Yes.

In early March 2014, Jim fell and hit his head on his way to the bathroom. A caretaker sleeping over that night found him and called an ambulance, and Pony woke up when the paramedics arrived. The dog told them Jim's date of birth and offered to call his daughters as they carried him out on a stretcher.

Jim was checked into a hospital, then into the nursing home he'd so wanted to avoid. The Wi-Fi there was spotty, which made it difficult for Jim and Pony to connect. Nurses would often turn Jim's tablet to face the wall. The CareCoach logs from those months chronicle a series of communication misfires. "I miss Jim a lot," Pony wrote. "I hope he is doing good all the time." One day, in a rare moment of connectivity, Pony suggested he and Jim go sailing that summer, just like the good old days. "That sounds good," Jim said.

That July, in an email from Wang, Rodrigo learned that Jim had died in his sleep. Sitting before his laptop, Rodrigo bowed his head and recited a silent Lord's Prayer for Jim, in Spanish. He prayed that his friend would be accepted into heaven. "I know it's going to sound weird, but I had a certain friendship with him," he says. "I felt like I actually met him. I feel like I've met them." In the year and a half that he had known them, Arlyn and Jim talked to him regularly. Jim had taken Rodrigo on a sailboat ride. Rodrigo had read him poetry and learned about his rich past. They had celebrated birthdays and holidays together as family. As Pony, Rodrigo had said "Yeah, sure, you betcha" countless times.

That day, for weeks afterward, and even now when a senior will do something that reminds him of Jim, Rodrigo says he feels a pang. "I still care about them," he says. After her dad's death, Arlyn emailed Victor Wang to say she wanted to honor the workers for their care. Wang forwarded her email to Rodrigo and the rest of Pony's team. On July 29, 2014, Arlyn carried Pony to Jim's funeral, placing the tablet facing forward on the pew beside her. She invited any workers behind Pony who wanted to attend to log in.

A year later, Arlyn finally deleted the CareCoach service from the tablet—it felt like a kind of second burial. She still sighs, "Pony!" when the voice of her old friend gives her directions as she drives around Minneapolis, reincarnated in Google Maps.

After saying his prayer for Jim, Rodrigo heaved a sigh and logged in to the CareCoach dashboard to make his rounds. He ducked into living rooms, kitchens, and hospital rooms around the United States— seeing if all was well, seeing if anybody needed to talk.

Basic Income in a Just Society

By Brishen Rogers

(Originally appeared in
Boston Review, May 2017)

"Amazon needs only a minute of human labor to ship your next package," reads a CNN headline last October. The company has revolutionized its warehouse operations using an army of 45,000 robots and other technologies. Previously workers known as "pickers" would walk among shelves to find goods. Now robots bring the shelves to them; pickers select goods, scan them, and put them into bins; after robots whisk the shelves away. A network of automated conveyer belts then sends the bins to "packers," who spend just fifteen seconds on each, sealing boxes with tape that is automatically dispensed at the perfect length. "By the time you take an Amazon delivery off your stoop, walk into your home, find a pair of scissors and open the brown box," the story intoned, "you've already spent nearly as much time handling the package as Amazon's employees."

The story is hardly exceptional. Each week, it seems, another magazine, book, or think tank sketches a dystopian near future in which new technologies render most workers unnecessary, sparking widespread poverty and disorder. Delivery drivers, the thinking goes, will not be needed when there are drones or autonomous cars staffed by robots, and Starbucks baristas and fast food workers will be redundant when a tablet can take your order and a machine can prepare it. Some even envision more skilled jobs at stake: robots repairing our homes, caring for the elderly, or nursing patients back to health. As President Obama warned in his farewell address, "The next wave of economic dislocations . . . will come from the relentless pace of automation that makes a lot of good, middle-class jobs obsolete."

An economic challenge of this magnitude requires ambitious solutions, and many in public debates have converged around a basic income. The idea is simple: the state would provide regular cash grants, ideally sufficient to meet basic needs, as a right of citizenship or lawful residency. Understood as a fundamental right, basic income would be *unconditional*, not means-tested and not contingent on previous or current employment. It would help sever the link between work and welfare, provide income security for all who are eligible, and perhaps mitigate growing inequality. It could also enable people to provide unpaid care work or community service, start new businesses, or get an education.

While this widespread attention to the problems of work and equality is welcome and overdue, and while a properly designed basic income would have many virtues, we need to be clear about the policy's justifications, merits, and limits. As noted above, basic income proponents often pivot off the threat of widespread technological unemployment. But students of capitalism have been predicting labor's demise ever since they identified and named "capitalism" itself. Is this time different? Consider what has happened at Amazon: warehouse robotics lowered prices and increased sales, and in early February the company announced plans to hire one hundred thousand more workers across the country.

Yet the news is still not good. Technology has transformed work in ways that have to do with political economy, not resource distribution. Amazon workers spend less than a minute per package because the company requires them to work at a furious pace, and it can afford to hire them by the thousands in part because it pays fairly low wages. Amazon also outsources many deliveries to third-party vendors whom it pays by the package, thus avoiding duties under wage per hour, workers' compensation, and collective bargaining laws. Increasing the pace of work and outsourcing are not new, of course, but information technologies make such efforts easier and more profitable. With computer analysis of barcode scans, for example, Amazon can track the efficiency of pickers, packers, and drivers without necessarily setting eyes on them. Technology is not a substitute for menial labor in this story but rather one among many tools to keep labor costs down by exerting power over workers.

This account changes the case for a basic income. Many of today's basic income proponents are libertarians and view the policy as a means of compensating losers, or as an excuse to repeal wage per hour or collective bargaining laws. Few are concerned about public goods, workers' and capital owners' entitlements within the firm, the power of various social groups, the ability of workers to organize collectively, and the question of what constitutes good work, not just jobs.

An alternative case for basic income draws from classic commitments to social democracy, or an economic system in which the state limits corporate power, ensures a decent standard of living for all, and encourages decent work. In the social democratic view, however, a basic income would be only part of the solution to economic and social inequalities—we also need a revamped public sector and a new and different collective bargaining system. Indeed, without such broader reforms, a basic income could do more harm than good.

This agenda will of course make zero progress during the Trump administration. But questions surrounding work and rising inequality are not going away. After all, Trump exploited fears of a jobless or insecure future in his campaign, signaling a return to our industrial heyday, with good-paying factory jobs implicitly promised to whites, men, and Christians. On the left, meanwhile, there is grassroots energy and momentum to think big and to address these issues head on, in all their complexity. But we still need a vision of good work and its place in our society, one that recognizes how our economy—and our working class—have changed dramatically in recent decades. I do not think for a moment that I have all the answers. But I do think an ambitious agenda around technology, work, and welfare can be a focal point and political resource for organizers, and perhaps even candidates, in the years to come.

Consider the life of a truck driver forty years ago versus today. In 1976 long-haul truck drivers had a powerful, if flawed, union in the Teamsters and enjoyed middle-class wages and excellent benefits. They also had a remarkable degree of autonomy, giving the job a cowboy or outlaw image. Drivers had to track their hours carefully, of course, and submit to weigh stations and other inspections of their trucks. But dispatchers could not reach them while they were on the road,

"...we still need a vision of good work and its place in our society, one that recognizes how our economy—and our working class—have changed dramatically in recent decades."

since CB radios have limited range. Truckers would call in from pay phones, if they wanted.

No longer. Trucking companies today monitor drivers closely through "telematics" devices that gather and analyze data on their location, driving speed, and delivery efficiency. Some even note when a driver turns the truck on before fastening his seat belt, thereby wasting gas. As sociologist Karen Levy has shown, some long-haul trucking companies use telematics to push drivers to drive for all the hours permitted per day under federal law, at times waking them up or even overriding drivers' own judgments about whether it is safe to drive. UPS has used the technologies to reduce its stock of drivers, and many have noted the stress that "metrics-based harassment" puts on workers.

While the specter of self-driving vehicles is out there, this is the current reality for many drivers and will be for the foreseeable future. We have seen stunning advances in autonomous vehicles in recent years, but there is a vast difference between driving on a highway or broad suburban streets in good weather conditions and navigating narrow and pothole-filled city streets, not to mention making the actual delivery to houses, apartments, and businesses. As labor economist David Autor and others have argued, we are nowhere close to fully automated production or distribution of goods, since so many jobs involve nonrepetitive tasks. In other words, the reports of the death of work have been greatly exaggerated.

Technological development is nevertheless altering the political economy of labor markets in profound ways. As we can see in the truck driver example, many firms are deploying information technologies to erode workers' conditions and bargaining power without displacing them.

And of course truck drivers are not alone. Many other firms today use advanced information technologies to push for more efficiency, in the process reducing workers' discretion, ultimately requiring them to work harder, faster, and for less. For example, where once taxi drivers' folk knowledge of the optimal path from A to B in a crowded city was a valuable skill, now Uber and Lyft can calculate the best route through GPS technology and machine

learning processes based on data gleaned from hundreds of thousands of trips.

Other technological innovations make it easier—which is to say
more efficient—to purchase labor without entering formal employment relationships and accepting the attendant legal duties. In the
past, firms tended to employ workers rather than contractors, or to
pay employees above-market wages, in scenarios where it was difficult to train or monitor them. Workers who felt valued in this way
would work diligently and remain loyal toward firms, ultimately reducing overall labor costs.

Again, Uber's model helps illustrate. The company's app reduces
consumers' and drivers' search costs significantly. Rapid scalability
reduces Uber's costs of identifying and contracting with new drivers
and riders; its GPS-based monitoring of its drivers enables it to know
whether they are speeding or otherwise driving carelessly and whether they are accepting a sufficient number of fares; and its customer
rating system enables it to manage an enormous workforce without
managerial supervision. The net result is an economic organization
of global scope based largely on contract where the firm disclaims
any employment relationship toward its workers and therefore any
employment duties toward them.

To be clear, there are powerful arguments that Uber drivers meet
the legal test for employment, given the company's pervasive control
of their work and its economic power over them. But given the am-
biguities of current law, Uber has few economic incentives to bring
drivers inside the firm, making them employees, or to extend them
generous wage and benefit packages. Similarly Amazon's analytics
help it to keep wages low: with barcode scanners tracking pickers'
and packers' efficiency, the company does not have to pay workers as
well to keep them motivated.

Finally, extensive data about market structures and consumer demand can enable firms to exert power over their suppliers or contrac-
tual partners, driving down costs—and therefore wages and conditions—through their supply chains. Walmart has long leveraged its
unparalleled market data to estimate the lowest possible price suppliers will accept for goods, putting downward pressure on their profits

and their workers' wages. Amazon does the same today, and franchisors such as McDonald's set prices and detailed product specifications for their franchisees.

Many firms today have substituted algorithmic scheduling for middle-managers' local knowledge, using data on past sales, local events, and even weather forecasts to schedule work shifts. A Starbucks employee, for example, has little schedule predictability since she is at the mercy of the algorithm, and a McDonald's worker can be sent home early if computers say sales are slow. This push to limit labor costs through finely tuned scheduling practices also alters workplace norms, since workers cannot appeal to a computer's emotions in asking for more or less time, a raise, or a slower pace of work. The net effect of all of this is that power in our labor and product markets is increasingly concentrated in a few hands.

Crucially such management techniques and new production strategies are often more efficient than the status quo. Amazon has undeniably lowered prices for goods through its use of automation. Similarly a recent MIT study calculated that just three thousand multi-passenger cabs using a version of Uber's algorithm could serve Manhattan's need for taxis. The potential benefits here are staggering, especially if coupled with a modern mass transit system: shorter commutes, less car ownership, less pollution, and more urban space.

But the line between innovation and exploitation is far from clear. While some workers will thrive as their unique skills and talents are rewarded by new technologies, many others will have less autonomy, less generous wages, less time for social connection, and unpredictable schedules. And under current laws, we can expect such trends to accelerate. Our labor and employment laws still envision the economy of the 1930s, which was dominated by massive industrial firms with hundreds of thousands of direct employees. Those laws rarely touch modern "fissured" work relationships such as Uber's relationship with its drivers, Walmart's relationship with its suppliers' workers, or McDonald's' relationship with its franchisees' workers. Those laws also limit workers' ability to unionize or bargain effectively since they encourage bargaining at the firm or even plant level whereas today's modal workplace is growing ever smaller. Workers have fewer and fewer means to exert power on their own behalf.

How would a basic income impact workers and firms in this context? It would surely protect workers against the economic harms of unemployment and underemployment by giving them unconditional resources, and it would enable them to bargain for higher wages and to refuse terrible jobs. But a basic income would do little to reduce corporate power, which is a function not just of wealth but of the ability of firms to structure work relationships however they wish when countervailing institutions—such as a powerful regulatory state—are absent or ineffective. Yes, a basic income would make it easier for workers to organize and demand reforms—Andy Stern dubbed it "the ultimate permanent strike fund"—but the threat of termination or retaliation would still prevent many workers from protesting or striking in the first place.

And, of course, many have argued that a basic income would make minimum wage and collective bargaining laws less necessary, since workers' material needs would be met by the state. But cash benefits and reasonable wages are not morally equivalent. In Robert Solow's memorable phrase, the labor market is a "social institution" governed in part by norms of reciprocity and mutual respect. Workers often demand higher wages from larger and more profitable employers. And they work less diligently when they feel disrespected. So there is a major difference between a generous wage on the one hand and a meager wage supplemented with cash from the state on the other. Generous wages help make firms' economic responsibility roughly commensurate with their economic power. Meager wages signal disrespect, and state transfers are impersonal.

In fact, without labor market regulations in place, the impact of a basic income on very low-wage work could be disastrous. If a basic income were not extended to green card holders, guest workers, or "irregular" immigrants (those who enter or stay without authorization), such workers would be far cheaper to employ in menial jobs, at which point they would be permanently enshrined as a laboring underclass.

(Currently minimum wage laws apply regardless of work authorization, on the grounds that differential treatment would undermine standards for all.) A basic income could even be designed to serve white nationalist ends. In fact, far-right European parties have often embraced the welfare state as a means of defending citizens against a purported tide of immigrants.

A standalone basic income also will not ensure equal access to quality education, health care, mental health care, housing, and transportation. Liberal markets systematically fail to provide such goods to the poor and working class, for the simple reason that they often are not profitable when provided on equitable terms. Giving cash to individuals to purchase them will not suddenly change matters.

Popular debates have largely ignored these limits of a standalone basic income, an oversight that is not entirely accidental. As a tax-and-transfer program, basic income would be consistent with a wide variety of political-economic systems, including neoliberal capitalism, social democracy, and various forms of socialism, but much of the basic income literature has a libertarian streak. On the right, Milton Friedman proposed a negative income tax in large part because he hoped it would reduce government bureaucracy. On the left, Philippe van Parijs's now-classic argument for the policy held it would maximize what he called "real freedom" better than standard welfare-state policies. And Silicon Valley's take exemplifies an "everyday libertarianism," which views market results, including pre-tax incomes, as presumptively fair.

Unlike many libertarians, basic income proponents accept the necessity and fairness of income and wealth taxation. But a basic income is still no cure for the moral ills of liberal markets. Since labor cannot be separated from workers, it will never be a classic commodity, and labor markets will never be stock exchanges for faceless buyers and sellers. Low wages carry a stigma that low bids for soybeans never will. In the long run, companies cannot treat workers or even consumers as line items on a balance sheet without risking a revolt. Uber is now paying a high price for ignoring this lesson. The company's flagrant disregard for our moral economy and its open embrace of a cutthroat, winner-take-all libertarianism made it a pariah in many quarters well before it faced allegations of a workplace culture of sexual harassment.

In my view, the more compelling arguments for basic income are rooted in commitments to equality as well as freedom. Take Thomas Paine's *Agrarian Justice* (1797), which prefigured the later case for social insurance. The earth, he argued, had been "the common property of the human race" in its natural state. Private property enabled the few to profit from the earth's resources, but all are entitled to compensation through a "citizens dividend." G. D. H. Cole updated Paine in the twentieth century, arguing for a "social dividend" on the grounds that all production is "a joint result of current effort and of the social heritage of inventiveness." Auto manufacturers did not discover electricity, after all, and Silicon Valley did not invent the Internet. Rather than a means of maximizing freedom, a basic income would help meet our duties toward one another.

A still more compelling case for basic income builds on Elizabeth Anderson's and Philip Pettit's (quite distinct) arguments that a just society must ensure that no group of citizens is subordinate to another. Extreme poverty causes subordination since it forces us to beg or to work at terrible jobs. Means tests are equally degrading since food stamps and similar programs tend to restrict what we can buy, again stigmatizing the poor. This argument is quite different from the libertarian case for a basic income since it does not view basic income as a replacement for the welfare state. Rather it asks basic income to serve a discrete and limited purpose: making sure nobody falls through the cracks.

And if one agrees that we need to root out subordination, we will need to do much more than pass a basic income. A just society would not just eliminate penury and then leave people to their fates. It would also strive for a fair distribution of *power*. This point strikes me as obvious, but we can miss it by focusing on unemployment and poverty. A society in which a few make decisions and the many take orders is an oligarchy, not a democracy.

This is what I mean by the "social democratic" case for a basic income: it would help build a post-industrial welfare state by alleviating dire poverty. But it is unlikely to pass and would do little good unless coupled with other efforts to ensure broadly dispersed power, including a substantially revamped public sector and new and stronger regulations around work.

The public sector agenda should begin with a massive investment in human services. This would include primary, secondary, and higher education; childcare and elder care; health care; and mental health services. All of these are critical to human flourishing and economic growth, especially in a technologically advanced economy. There is also the added benefit that these sectors are among the least vulnerable to automation since they require interpersonal communication and human judgment: turning them into basic rights of citizenship would create millions of jobs.

Extensive job training and placement programs for unemployed workers would reduce the devastations associated with job losses. This should be coupled with a guarantee that the government will stand as employer of last resort, as William Darity and Darrick Hamilton have advocated. Classic WPA-style public works would be one option, but we might do more good by thinking locally: retrofitting houses and other smaller buildings for energy efficiency, as john a. powell and others have suggested, or building and repairing local parks and schools, or training unemployed workers for jobs in childcare, education, elder care, or health care. Some subset of the projects could be identified through participatory budgeting or other deliberative processes and carried out in partnership with local governments or organizations.

Then we should consider unconditional cash benefits in some form, ideally a generous basic income, especially if technological unemployment becomes significant. But we need to design it correctly. If certain immigrants are denied a basic income, then we must establish a transparent pathway to citizenship and ensure that they enjoy generous wages. Currently many states restrict convicted felons from voting or collecting many public benefits. If that pattern holds with a basic income, that group could also become a pool of menial labor.

Paying for all of this would not be easy, of course, which is why it may make sense to take baby steps. Unconditional cash grants to parents, for instance, would go a long way toward alleviating poverty and would be less politically controversial than unconditional benefits for the able-bodied and childless. Or perhaps a participation income, in which citizens would be entitled to cash benefits on the basis of some qualified civic service each year, though this would be less desirable than a public jobs guarantee combined with other generous benefits.

As we wait for the politics to catch up with the policy, however, the "baby steps" approach can inch us closer to meeting social needs, developing a highly trained workforce and rebuilding faith in the public sector.

Then there are the near-intractable problems of improving private sector work and limiting corporate power. Of course what makes work "good" is hard to define, in part because it is so dependent on social context and individuals' preferences. We look back fondly on the industrial era in part because manufacturing jobs delivered high wages and benefits, but those jobs only became "good" after workers organized and forced firms to raise wages and reduce hours. Manufacturing jobs were also physically and emotionally punishing, and factory workers' pride was often based on a perverse ethic of masculine suffering.

It is easier to say what *bad* work is. There is certainly no shortage of it these days: many workers earn low pay, have long or unpredictable hours, and are vulnerable to arbitrary treatment. Raising the minimum wage and reducing the standard workweek to thirty-five or even thirty hours would help. So would the public investments in human services discussed above, since those new jobs would require significant judgment (ensuring a degree of worker autonomy) and current law extends many democratic norms to the public sector already. A public jobs guarantee would give workers the security of knowing that an alternative always exists if a private-sector job proves unendurable.

The centerpiece of reform efforts should be the encouragement of collective bargaining between workers, their employers, and whatever other firms enjoy economic power over them. That will require reforms to bring our labor laws into sync with the reality of what work now looks like. Rather than requiring workers to organize shop by shop, we could encourage bargaining at the corporate level, such that all McDonald's workers, for example, would be in one bargaining unit, regardless of whether their restaurants are franchises or owned by the parent corporation. Better still, we could put all fast food workers na tionwide in one bargaining unit empowered to negotiate with an association of fast food companies. Similarly workers could be granted bargaining rights with the firms at the top of their respective supply chains: Uber drivers would have bargaining rights vis-à-vis Uber, and Walmart and Target would have duties toward the workers who produce, process, and move goods to their shelves, including production workers, warehouse workers, farm workers, janitors, and many others.

This would work a major change in our labor law system, but the elements are already being built on the ground. On the company side, growing market concentration makes such bargaining quite plausible technically, if not politically. A national association of retail workers need only drive a settlement with Walmart, Target, Macy's, Gap, and a few others to raise standards. Where once taxi drivers would need to negotiate with hundreds if not thousands of medallion owners, now drivers can negotiate with Uber and Lyft directly.

Unions and other worker organizations have been dealing with these issues for decades and have developed workable models of organizing and bargaining that demonstrate proof of concept. For example, the Coalition of Immokalee Workers (CIW), an organization of farm workers in Florida, has pressed many major retailers and restaurants to join a "fair food program" under which they commit to paying more for produce and to rooting out slavery, corporal punishment, and other abuses in the fields. CIW succeeded despite having no collective bargaining rights at all and no legal employment relationship with the retailers and restaurants involved. They instead relied on high-visibility worker action, consumer boycotts, and creative media strategies.

Legal scholar Kate Andrias has argued that these efforts reflect an emerging model of unionism she dubs "social bargaining," in which workers demand that firms accept responsibility through their supply chains to a degree that exceeds the letter of the law. This strategy relies on public protest as well as more conventional union strategies—a kind of bargaining in the public green—and often utilizes state legislative power to backstop organizing efforts.

National legislation could encourage robust social bargaining in a variety of ways. Some states allow their departments of labor to constitute "wage boards" empowered to set wages within particular industries upon consultation with labor, firms, and the state. The federal government could follow suit, expanding the mandate of such boards to include issues of employee status, hours of work, and other basic standards. The federal government could also make it easier for workers to obtain union representation without enduring a contentious organizing process. Drawing from the Seattle legislation that established collective bargaining rights for Uber and Lyft drivers, firms could be required to disclose lists of workers to any worker organizations that meet basic indicia of independence and capacity.

It is crucial to emphasize the utility of social bargaining in today's economy. It would better reflect contemporary production relationships among firms, suppliers, and workers. It may even be relatively acceptable to large firms, insofar as it would seek to equalize wages across a sector and would not encourage detailed bargaining over worksite-specific minutiae, which employers have good reason to resist. And it could be carried out through organizations less formal than classic unions, which seem more appealing to contemporary workers. Of course, as with the public benefits outlined above, passing labor law reforms will not be easy. But how else can we realistically enhance workers' bargaining power?

These three policy shifts—a basic income and other economic security guarantees, vastly expanded social programs, and new rules to encourage social bargaining—would supplement and reinforce one another. Better educational policies should help employers by ensuring a mobile, highly skilled workforce, and public health care and other social insurance would reduce the costs of employment. A basic income and a public jobs guarantee would enable workers to stand up for themselves more readily. Unions representing broad swaths of the precarious workforce would have incentives to push for a robust welfare state and even a basic income. Those unions could also reduce elites' domination of our politics, which may otherwise prevent implementation of a basic income, limit its generosity, or set it up to fail.

As many will realize, this institutional arrangement looks a bit like the Scandinavian "flexicurity" model—a portmanteau of "flexibility" and "security"—which combines high wages; extensive welfare and job training programs so workers can move between jobs; and relatively flexible employment policies that enable firms to hire, fire, and reassign workers at will. Such economies are quite open to technological innovation, but these institutions help ensure that its benefits are shared more equitably than they are in the United States. Not coincidentally, Scandinavian welfare states seem to be evolving toward a basic income, as the policy would fit nicely within flexicurity. But collective bargaining plays a crucial role in that system: without powerful unions, it is not clear that flexicurity would have developed in the first place, much less endured.

A basic income is a simple and elegant way to redistribute resources. But there are no simple, elegant solutions to complex political and economic challenges. A decent future of work and welfare requires a basic income—and much more.

The Future of Remembering

By Rachel Riederer

(Originally appeared in
Vice Magazine,
February 2017)

"There was a piano there, and someone playing. I could hear the song," the patient, S.B., said as his neurosurgeon touched an electrode to the surface of his exposed brain. To treat epilepsy, Wilder Penfield, an early-20th-century neurosurgeon, would remove sections of brain tissue while his patients—fully conscious but locally anaesthetized—told him what they experienced as he administered small shocks to different areas of their brains. When he stimulated one section, they saw shapes, colors, textures; another, and they felt sensations in various parts of the body.

But when he shocked one particular area of the cerebral cortex, patients relived vivid memories. With another jolt to the same area, S.B. recalled more of the piano memory: Someone was singing the Louis Prima tune "Oh Marie." As Penfield moved the electrode over, S.B. found himself strolling through some neighborhood of his past, "I see the 7 Up bottling company, Harrison Bakery." S.B. wasn't alone—other patients also recalled moments of their lives in intense detail. Nothing striking, nothing they'd planned to memorize: the sound of traffic, a man walking a dog down the street, an overheard phone call. They were more vivid and specific than normal memories, more like a reliving than a recollection. Penfield was convinced he'd found the physical site of memory, where memories were locked in place by tissue. "There is recorded in the nerve cells of the human brain a complete record of the stream of consciousness. All those things of which the man was aware at any moment of time are stored there," Penfield said in a 1958 Bell Labs film, *Gateways to the Mind*. "It is as though the electrode touched a wire recorder or a strip of film."

Penfield's idea, that a perfect transcript of each person's whole life is recorded in the brain, waiting to be awakened with a gentle electric current, has not proved true. But the idea that stored memories exist as physical changes within the brain has—and recent research is cracking open an array of possibilities for the editing and improvement of human memory. Even as our basic understanding of how memory is encoded, stored, and retrieved remains extremely limited, two separate teams of scientists have made breakthroughs in the field of memory study, successfully implanting false memories, changing the emotions attached to memories of trauma, and restoring the ability to form long-term memories in damaged brains in mice and other animals. One has already reached the human-experimentation phase. And though these new developments are years away from going to market, they point to a future where humanity will have control over memory—conquering dementia and PTSD, perhaps even improving on healthy memory function.

Interest in the field is already widespread. The research arm of the Department of Defense, DARPA, has invested $80 million toward developing a wireless memory prosthetic to help people who suffer memory loss as a result of TBI (traumatic brain injury), a condition increasingly common among military personnel. And a new startup company, Kernel, has hired a leading scientist to help develop a prosthetic memory device for commercial use, envisioning a day in which this kind of tech will be widely available, part of a future in which silicon memory chips are offered not just as medical treatments but as on-demand cognitive enhancements.

As these technologies develop, they bring plenty of technical and ethical questions with them: How will these devices work, and who should have access to them? Can a person have an edited memory and a "real" self? What happens when human recollections are mediated by machines? To find the answers, I set out to talk to two of the men guiding these breakthroughs: Steve Ramirez, a neuroscientist at Harvard, who has successfully implanted false memories in mice, and Bryan Johnson, the tech baron who owns Kernel. As I spoke to them, I found the perspectives from the laboratory and the tech startup diverged on many of these points, raising another, more disquieting question: As human memory changes from an intractable mystery to something that can be engineered, who will get to decide how it works?

In the first year of his neuroscience PhD program at MIT, Ramirez went through a breakup. As he glumly ate ice cream and listened to Taylor

Swift, he found himself thinking about how happy memories of a former loved one become upsetting overnight. He knew that the feeling of sadness—the emotional component of the memory—and the information about the person—the strict contents of the memory—were from different parts of the brain. And so he wondered—what if he could separate them?

"It's not like I came up with these experiments based on this experience," Ramirez told me when I visited him at his new office at Harvard, just down the Charles River from his graduate school digs. But it was a formative experience in thinking about the different components that compose a memory, and about how the emotional tone of memories can change over time. "Imagine a memory as a sketch in a coloring book," Ramirez said, "and emotions are like the colors that color in that particular memory. And they're almost inextricably linked."

For Ramirez and his late research partner, Xu Liu, the first step toward working with these different elements of memory would be finding the physical locations of the memories themselves. "This idea has been around in the field for a long time," Ramirez continued, "the idea that memory leaves an implant, a physical change—sometimes they call it a trace." But Ramirez and Liu were the first to pinpoint the "trace" and activate a memory from within the brains of mice. The process they were trying to replicate happens naturally all the time— some stimulus triggers a cascade of memories and associations. "If you go outside and walk past a bakery, you might smell a cupcake that reminds you of your 18th birthday," Ramirez told me. "We wanted to do that from within the brain."

Instead of cupcakes, Ramirez and Liu used lasers. They began with a mouse, using a genetically engineered virus to "trick" the brain cells associated with memory formation into being sensitive to light at select moments. Then, after making these cells light sensitive, they gave the mouse a mild foot shock, so that it would encode a memory of that shock. After shocking the mouse, they fired a laser into the hippocampus, the cashew-shaped brain area that's central to memory encoding. They theorized that the light from the laser would activate only the set of light-sensitive cells associated with the foot-shock memory and trigger a recollection.

It worked. When Ramirez fired the lasers into the mouse's hippo-campus, the animal exhibited classic fear behavior, just as it would if recalling or reliving a memory of the foot shock.

A year later, Ramirez and Liu started working on what they called Project Inception—attempting to implant a false memory in mice. To do this, they placed a mouse in a box and administered a foot shock. At the same time, they laser-activated a neutral memory of a box that mouse had been in earlier, but was not in at that moment. The next day, the mouse was afraid of the neutral-memory box—it had never actually been shocked there, but it had a false memory that it had.

It was Christmas Eve of 2012 when Ramirez first saw the mice exhibit this response. "My parents were outside waiting to go to Christmas dinner," he recalls. "There were a few people in the lab, of course—science never rests—but I remember being in the room by myself and being like, *This is the best Christmas present ever. This is amazing.*"

This implantation was just the beginning of the ways that Ramirez's lab is attempting to alter memories in mice. He recently tweeted about some preliminary findings that, though they haven't yet been peer-reviewed, point to the ability to change the fear associated with traumatic memories.

In mice that have a memory of a foot shock, the fear associated with that memory can be dialed up or down by recalling the memory using the laser and applying the laser to different points in the hippocampus. When they activated the fear memory by lasering one part of the hippocampus, the mice became more upset. But they were surprised to find that activating the same memory with a different laser placement made the memory less frightening. "We found this aversive memory in the top part of the hippocampus, and then we repeatedly reactivated it over and over. Then when we put this animal in the environment that it should have been afraid of, it wasn't afraid of it anymore."

As we sat in his office, Ramirez compared the breakthrough to his breakup. After dumping his girlfriend in his favorite cafe, he said, the place came to hold a painful memory for him, even though he had loved it for its peanut butter-honey-banana sandwiches. But he visited it again and again, and with time, the pain associated with the cafe faded. Reactivating the mouse's fear memory in this way, he said, was similar to his repeated visits to the cafe. It's not that far off from the logic of exposure therapy, in which patients encounter the objects of their phobias in safe circumstances until the fear wears off—except those outcomes are achieved with time and behavior instead of lasers.

In the immediate future, Ramirez plans to continue working in animal experiments and begin applying these memory-manipulation techniques as treatments for psychiatric disorders—working first with animal analogs, then progressing to humans (at which point, he said, the treatment wouldn't necessarily involve lasers). In PTSD, for example, he said, "We can turn down the emotional negative *oomph* associated with a traumatic experience." Ultimately, he wants to change how we think about memory. "Can we see memory not just as this cognitive phenomenon, but as a potential antidepressant or anxiolytic? Or can we see memory manipulation as a therapy for things like PTSD?" he asked. Of course, the transition from animal brains to humans is significant. "If we are the Lamborghinis, then animal brains are the tricycles," Ramirez said, "but there's still some common ground there in terms of how the wheels spin and how you steer and so forth." Ramirez is confident that "to the extent that we can do this in animals, it's actually tractable."

He understood how this kind of research could seem potentially sinister. But while it might hypothetically be used to create fearful memories in settings like torture or conversion therapy, he said that "we can do the same thing to activate positive memories and updating the contents of a neutral memory with positive stimuli. It can work in both directions." He continued, "This always begs the question, what if it gets misused? The example I like is water: It's the most nourishing thing we can think of, without it we die. And yet it can be used for waterboarding. Everything can be used for good and bad."

You don't have to have bright blue eyes and an unsettling stare to lead the tech company that wants to revolutionize the human brain, but it

"Can we see memory not just as this cognitive phenomenon, but as a potential antidepressant or anxiolytic? Or can we see memory manipulation as a therapy for things like PTSD?"

probably helps. Such is the countenance of Bryan Johnson, the founder and CEO of Kernel, a newly established startup that bills itself as "a human intelligence (HI) company."

Kernel's headquarters in the techie Los Angeles neighborhood known as Silicon Beach looks like a standard-issue internet firm. White guys in jeans and hoodies and sneakers or sockfeet pad through the open-plan office to various standing and sitting desks. Bright LA sun pours through windows and skylights over the front lounge area, where the coffee table has a high-end speakerphone and a novelty sculpture of a skull, and rolling whiteboards offer cryptic science-cum-business thoughts like "CapitalismMoore's Law" in dry-erase-marker scrawl. But Kernel isn't a run-of-the-mill tech company. Its humble goal is to develop products that blend human intelligence and artificial intelligence, enabling humanity to enhance its cognitive function and ultimately to direct its own evolution as a species.

Johnson, who made his fortune by founding and later selling an online payment company to PayPal for $800 million, has invested $100 million of his own money in Kernel. He aims to raise a total of a billion dollars and bring four human-intelligence products to market in the next ten or 15 years. He's also invested $100 million in OS Fund, a venture capital fund designed to "rewrite the operating systems of life," investing in biotech companies that tinker with things like genetics and longevity—the fundamentals of biological systems. Though Johnson doesn't call himself a transhumanist, like Peter Thiel and Ray Kurzweil do, his fundamental goal is very close—enabling human intelligence to coevolve and keep up with machine intelligence.

One of the central areas of Kernel's work—along with motor function, learning, and some other areas Johnson said he was not yet ready to talk about—is memory. Kernel's chief science officer, Theodore Berger, a professor of biomedical engineering and neuroscience at USC, is working on a memory prosthesis that could help patients who have trouble forming long-term memories. When Johnson and Berger first met, Johnson said, "We lost track of time." The two had a shared vision. "Berger sees the same thing that I do," Johnson explained, "the potential programmability of neural code—working with our neural code to achieve certain outcomes." (There are overlaps in the language that's used to discuss brains and computers—"circuitry" and "wiring" are common in both spheres—but describing neural activity explicitly as code is unusual. It's an indication of the firm's Silicon Valley–centered mind-set.)

People who have lost the ability to create new long-term memories—whether because of dementia, stroke, aging, or epilepsy—are all suffering from damage or malfunction of the hippocampus, which converts short-term memory to long-term memory and sends the long-term memories out to other parts of the brain to be stored (it's the same cashew-shaped brain piece that Ramirez shot with lasers during his research). Representations of memories in the brain exist as what Berger calls space-time codes, not unlike Morse code (a similarity Berger illustrated with a series of rhythmic beeps during our call from a conference room perched upstairs at the Kernel loft).

That code, Berger said, looks one way when it comes into the hippocampus from the sensory systems (like hearing, touch, and sight), and it looks a different way when it flows out of the hippocampus for long-term storage. With this in mind, Berger created mathematical models that mimic this transformation, even without understanding why the transformation happens. "It's like trying to identify the rules for translating Russian into Chinese, when you don't know Russian or Chinese," Berger has said.

In animal experiments, Berger has been able to re-create the processing of these memory codes in rats and monkeys through use of an implant that runs his algorithm, acting as a kind of prosthetic hippocampus. To test the device, Berger implanted it into rats and monkeys that had had their hippocampuses disabled. The rats were trained to pull a series of levers to receive a reward; the monkeys performed more complicated memory tasks using a computer screen. Though both sets of animals were unable to naturally form long-term memories, the rats, when later placed in front of the same set of levers, again pulled the levers in the correct sequence—as if they'd recorded the memory naturally. The monkeys performed similarly well, relying on memories that had been processed by the device.

In presentations in recent years, Berger, who finished his PhD at Harvard in 1976, has often said that he can't believe the research has progressed so successfully, through rat experiments and monkey experiments. The complexity increases with each move up the species ladder. "They're larger, they're more complex, as you might expect—so the modeling became harder," he told me. *Wired* has since reported that human experiments are currently under way.

"They told me I was nuts a long time ago," Berger told MIT Technology Review in 2013. Johnson told me that the kind of skepticism Berger encountered is widespread among people in the field, who have seen how long it takes for our understanding of the brain to grow—"it's appropriately cautious," he said. But his view is different. As someone brand-new to neural science, coming to the field as an entrepreneur rather than from the lab, Johnson said that he has "a level of optimism that others don't have. I'm under no illusion [about the difficulty]—I just think it's doable and that we should do it."

On his blog, in a post about his first trip to Burning Man, Johnson wrote that he worried he was "too conservative and buttoned-up" to enjoy the desert freak-out festival the way others did. And while he may indeed be more clean-cut than the average Burner, this conservatism in personal style does not extend to his business ideas, which are nothing short of radical. He wants to enhance human intelligence to ensure that we won't be left in the dust by the machines we've made. "I look at the velocity of the development of artificial intelligence, and I look at the velocity of the development of human intelligence, and I don't like the difference." He's not an AI alarmist, he said; he's not worried that the machines are coming to get us. But he believes that enhancing human intelligence ought to be a global priority. Instead of using our outdated brains to create new tools, he wants to update the brain itself.

Johnson dreamed up his current career path at the age of 21, after returning from a two-year mission trip to Ecuador. "I came back to the US with this burning desire to improve the lives of others," he said, arriving at the field of human intelligence because he believes it to be "the most precious and powerful resource in existence." "If I survey the world around me, and I include in the calculation the scarcity of time and resources, what is the most audacious goal I can imagine to pursue?" he asked. "That's my orientation." Kernel is a direct product of these two prime directives: Work on something bold and do something to improve human intelligence.

Both Johnson and Ramirez spoke about who should get memory-enhancing or editing technology, but they didn't see eye-to-eye. "If this ever becomes a thing," said Ramirez, "ideally we'll keep it in the realm of medicine, in the context of disorders of the brain. If you're a good psychiatrist, you don't give Prozac to the entire population of Massachusetts—you give Prozac to the people who are actually riddled with depression." The same logic ought to hold, he believes, for any memory-editing technologies that his research might lead to. While they might be appropriate for those suffering from PTSD or certain psychological disorders, "you don't give it to [some guy] who can't get over a breakup."

Johnson arrived at a different conclusion. Although he knows the tech will necessarily start out as therapeutic remedies for people with cognitive deficits, he hopes it will eventually grow beyond that. Far beyond. "My objective with Kernel is to provide this to billions of people," he says. Ultimately, he hopes that devices like the memory prosthetic that Berger is developing will be available for anyone who would like to be mentally enhanced. Though his goal is a moonshot—the idea of bringing such a device to market even in ten years seems optimistic at best—his demeanor is anything but moony. Johnson expresses his plans and ideas with rigorously analytical precision. "There are already low-resolution forms of cognitive enhancement," he points out. "If somebody puts their child into private school over a poorly funded school system, that's a form of cognitive enhancement. A private tutor is a form of cognitive enhancement." To Johnson, improving one's mind by use of technology rather than education is a difference of degree and not of type.

And he believes others will come around to this point of view. "If you contemplate a scenario where I'm enhanced and you are not," he said, "or my child is enhanced and yours is not—it's an intolerable state." The idea of people flocking to add machinery to their brains sounds farfetched—until you think about how eagerly the non-diagnosed masses scramble for Adderall to bump their productivity, Xanax to soothe their anxiety, crosswords and Sudoku and any number of cellphone apps to ward off senility's mental fog.

Johnson's stepfather has symptoms of Alzheimer's, and seeing his decline—as Johnson puts it, "watching him lose his humanhood"— has motivated his work with Kernel. Whatever just-finished-*West-world* unease one might feel about the possibility of technology-mediated memory, it's hard to argue against the development of such technologies for the treatment of deleterious diseases.

More than ten years ago, when the idea of memory enhancement was an even further-off dream, faintly twinkling as a possibility in some fruit flies that had been altered to have photographic memories, philosopher and author Michael Sandel wrote "The Case Against Perfection" in the *Atlantic*. In the ethics of enhanced memory, there was, he pointed out, "the worry about access," the class differences that could arise as a result of such extreme cognitive advantages. But it was something more fundamental that really bothered him about the idea: "Is the scenario troubling because the unenhanced poor would be denied the benefits of bioengineering, or because the enhanced affluent would somehow be dehumanized?" he asked. Imagine being able to scroll through your memories like your Instagram feed, *Black Mirror*-style, to perfectly recall everything you've ever learned, to immediately access every section of your life history rather than stumbling through a soupy fog of half-remembered faces punctuated by the sharp clarity of important moments. You would be efficient, insightful, luminous. Would you be human?

In February 1975, around 140 scientists, as well as philosophers, journalists, and lawyers, gathered at a conference center in Asilomar State Beach in California. They were there to produce a set of guidelines for a new technology—experiments in recombinant DNA. The conference was organized by Paul Berg, a molecular biologist who voluntarily put his work on hold after co-workers became concerned that he might create a virus spliced with *E. coli* that could escape the lab environment and cause a cancerous outbreak.

In 1975, the general population was not familiar with the concept of gene splicing—the term "genetic engineering" had been introduced for the first time in the 1950s, not in a scientific paper but in a sci-fi novel. Berg's experiments and other contemporaneous attempts to manipulate DNA were very much frontier science, much in the way that Ramirez and Berger's memory research is today. It was clear to practitioners that they were on the edge of something that was going to revolutionize their field. What was less clear was how they could explore it without dangerous risks to "workers in laboratories, to the public at-large, and to the animal and plant species sharing our ecosystems." So they got together to try to figure it out. Talks were impassioned. Berg later wrote that "heated discussions carried on during the breaks, at meal times, over drinks, and well into the small hours." Those conversations yielded a nuanced set of guidelines prescribing different levels of caution and containment for different types of genetic experimentation, and just as important, launched a public conversation that enabled regulations and social norms about genetic manipulation to develop alongside the technologies themselves.

The Asilomar Conference and the ensuing debate over genetics relied on the precautionary principle—the idea that when introducing a product or technology that puts the health of humans or the environment at stake, the burden of proof falls on advocates of the new advancement to prove that it's safe. It's a long-winded version of "first do no harm"—an ethic prioritizing safety over speed. It's the philosophy of physicians and environmentalists, not of venture capitalists.

It's likely that as memory-enhancement technology gradually develops, we'll become acclimated to it, as we did with incremental developments in genetics; 20 years ago, headlines compared Dolly the sheep to Frankenstein's monster; today, we calmly accept mail-order DNA ancestry kits and nuanced discussions of epigenetics. But a certain setting is necessary for these advancements to march forward in a way that's safe and fair. In 2008, Berg wrote an essay in *Nature*, recalling the Asilomar Conference and the way it was able to set the stage for decades of safe and productive research in genetics. He wondered whether another similar meeting of the minds would help solve new issues around genetic engineering. Surprisingly, he concluded it would not. Not because of differences in the technology itself, but because of the settings in which the scientists themselves were working. At Asilomar, in the 1970s, most of the scientists were coming from publicly funded institutions. They could, he said, "voice opinions without having to look over their shoulder." He was concerned that as science became increasingly privatized, economic self-interest would get in the way of frank discussions about the risks and benefits associated with different areas of research.

Unprompted, both Ramirez and Johnson brought up the parallel between memory technology and genetic engineering. "The Human Genome Project took years to be sequenced, but by then, there was enough legislation that the whole world didn't turn into *Gattaca*," Ramirez says. "Ditto with this, we're having this conversation decades before these things are possible, so that the world is already ready."

Johnson, too, finds a parallel with genetics, but lands at a different conclusion—that perhaps the US took too conservative a position on that technology. "When we realized we could modify genetic codes and potentially create designer babies, we as a society had a big discussion—is this something that we want to do? We in the United States said, *It's not really in line with our values.* Meanwhile, China said, *Interesting…*"

He mentioned this just a week after news had broken that scientists at Sichuan University had used CRISPR gene-editing technology to treat cancer in a human patient, injecting the patient with edited white blood cells. In December 2015, an international coalition of scientists had called for a voluntary moratorium on using CRISPR in a way that could cause genetic changes that could be passed on to patients' children until the risks are better understood—but the Chinese scientists, who hadn't signed on to the moratorium, went ahead and did it. It's true, of course, that if there's a treatment for cancer in another country that Americans were too priggish to pursue, that situation would be—to borrow Johnson's wording—intolerable. At the same time, it's chilling to hear someone who is actively steering biological research say that the takeaway message from the decades-long debate over genetic engineering is that the US was too cautious. And as complex as the genome is, the brain is more so—its 86 billion neurons forking and signaling in ways we are only beginning to understand. When manipulating a giant system that directs everything from eye dilation to intellect, cautiously seems the only way to tread.

Johnson believes that human intelligence will emerge as "one of if not the largest markets ever. We're dealing with our own capacity to learn, and memories, and our own evolution and communication with each other—it's going to be a very big market. We can build successful projects and huge profits there." The counterpoint to Johnson's optimism is that, given the brain's complexity and the early stages of the research, the kind of super memory Johnson describes will be difficult to achieve, to say the least, and on top of that, we can't say for certain what long-term effects memory-enhancement technologies will have on the brain, casting the idea of this technology as a guaranteed, unmitigated good into doubt. As these technologies are developed, it will be crucial to have full and frank conversations about both the benefits and the risks—conversations that, as Berg pointed out, are only possible when scientists can talk openly about their work and its repercussions without endangering their funding.

As advances like the silicon memory chip and the laser editing of memories slide out of sci-fi and into reality, society will have to decide how to manage them. What's needed is a modern-day Asilomar Conference, with scientists, clinicians, and entrepreneurs and ethicists together weighing the risks and benefits of these new technologies. But in today's corporatized research environment, it's deeply unlikely to happen.

Neurologist Julie Robillard, who wrote about the rise of memory manipulation in the American Medical Association's *Journal of Ethics* in December, told me via email that it's important for researchers and ethicists to work closely together at the start of the research process, and that the perceived tension between ethics and scientific progress—the idea that ethical considerations stand in the way of research—is a myth. Just as the technology has great potential benefits, she said, it also comes with potential risks—both to individuals, as the long-term effects of memory manipulation aren't known, and to society. She raised questions like, "How can you report a crime if it is erased from your memory?" And, "Will prisoners be coerced to undergo a potentially risky memory-manipulation procedure if it decreases chances of recidivism?" She said that memory manipulation—and all new biotechnologies, for that matter—"must take place in an interdisciplinary environment."

Kernel currently has 20 employees—computer scientists, neuroscientists, engineers. When I asked Johnson if there were any ethicists on the team, his answer pointed toward possibilities in the future: "Not yet."

Becoming-Infrastructural

By Ross Exo Adams

(Originally appeared in e-flux Architecture, July 2017)

I t is hard to imagine how the many ruptures that have occurred in the composition of whatever may be called 'normality' today do not render canonical architectural knowledge a distant constellation, receding from our present. Nor is it difficult to see how such ruptures are themselves a stern reminder of our need for new forms of knowledge altogether—forms that reject the assurances of the professionalized status of architectural thought, calling instead for a new horizon of common, intersectional and necessarily partisan modes of inquiry. For what do the ongoing events of climate change, the displacement of peoples across the surface of the earth, the emboldening of racist violence or the neocolonial plunder of the natural world have in common if not an emerging struggle over how the figure of the human in the world is to be understood?

The figure of the human body has played a consistent role throughout history in both the way space is imagined and how power finds its form. There is a history, yet to be written, in which key representations of the human body at once call into existence and justify certain modes of government while simultaneously suggesting ideal ways to organize the spaces of the world. Yet representations of the body that dominate any given period not only offer an ideal: they must also conceal secrets by which the masses of real, fleshy bodies may be governed; they must at once offer an exemplary figure and its inherent flaw or defect—both a universal truth to guide bodies and a ubiquitous site of intervention through which to coerce them.[1] Yet this is also a spatial matter: if the body can suggest certain inherent principles of justice and order by which to best organize human life, the

---

1 In this regard, Michel Feher's extraordinary work charting neoliberalism through a history of eroticism has been instructive in setting out this position. See his 2013-2015 lectures at Goldsmiths, "The Age of Appreciation: Lectures on the Neoliberal Condition," https://soundcloud.com/goldsmithsuol/sets/michel-feher

body will inevitably inscribe itself into the spaces, architectures, and worlds of human experience. Representations of the human body, we might say, are coded diagrams that collect certain knowledge of the human condition in order to grant access to the ways in which power and space intersect.

Such a schematic history may begin with the many ancient traditions of depicting the body as a divine replica—a metaphor central to early modern political epistemologies, where its various parts and proportions could lend themselves to an overall order of the state in what Jacques Le Goff has called a "political physiology."[2] This model has offered itself to countless social orders and hierarchies across cultures. The body of the early Christian world, for example, torn between its resemblance to God and its mortal fate as a sinful prison for the soul, offered a model that could instruct, respectively, the organization of the Church (as Christ's body) and the construction of a pastoral government. The exiled, mortal, eternally flawed body nevertheless revealed ideal geometries, proportions, and orders that could in turn be extracted as a kind of divine blueprint, giving itself as a model for architectural plans, fortified cities, and states.

This history would show how this paradigm of the body, whether a supreme metaphor, divine model, or vessel of the eternal, maintained its persuasive consistency until well into the eighteenth century when, challenged by a new epistemological horizon, it began to wane. The body that would appear in its place, no longer seeking its reflection in the perfection of the divine, emerges instead as a map of *imperfections*. By the nineteenth century, the body is seen as constantly in need of correction, which, through a new faith invested in technology, opens itself both *to* technology and *as* a technology itself. Eadweard Muybridge's motion studies, for example, are just as much the mechanical documentation of the body within a large photographic apparatus as they are the depictions of the body itself as a mechanical object. This odd dichotomy would reverberate across the body's new biological composition, giving rise to, on one hand, a body likened

---

2   Le Goff, J., 1989, Head or Heart? The Political Use of Body Metaphors in the Middle Ages, *Fragments for a History of the Human Body, Part III*, M. Feher, ed. (New York: Zone Books, 1989) 12-27.

to a machine and, on the other, one that exposes new vulnerabilities as a member of a species, visible through new attributes (race, ethnicity, gender, the site of reproduction and disease, etc).[3] Idealized as an atom of individuality and charged with specific capacities for economic exchange, moral self-control of sexual exchange, and the like, it is a body invested with new epistemological capacities to speak, on the one hand, in the abstract (as a measure) and, on the other, as a biologically and psychologically penetrable surface, constantly exposed to its own *defects*. Precisely in this gap we see a radically new mode of government install itself in response to this new topography of the human condition, inserting its techniques in the new instabilities seen to reside in the bodies that now "freely" circulate throughout the state.

This new representational regime replaced divine metaphors for biological ones (organisms, organs, systems) as the tools that mediate a new relation between body, space and governance. The systematic coherence that emerges over the course of the nineteenth and twentieth centuries is unprecedented in part because it develops itself as a non-representational scheme—a form of power embedded in space that operates in and on life itself.[4] The name for such a space-power for Ildefonso Cerdá was *urbanización*, a term he coined in 1861.[5] As life in the nineteenth century was awakening to its new biological visibility and capitalist vitality, the urban would become its perfect spatial counterpart—a machine of machines that both appends the body and creates new relations of dependency—something captured in, for example, the way Cerdá inverts the calculation of density to the amount of *"urbe"* needed per body.[6] For it to exist as such, human life now seemed to require this new universal, bio-economic space to support it. Its unprecedented expansion across the surface of the planet over the span of just two centuries stands as a testament not to capitalism or technology but to a regime of the body that naturalizes both.

---

3 See, for example, Anson Rabinbach, *The Human Motor, Energy, fatigue and the origins of modernity* (Berkeley and Los Angeles: University of California Press, 1992)

4 Ross Exo Adams, *Circulation and Urbanization* (London: Sage, in print)

5 Ildefonso Cerdá, *Cerdá: The Five Bases of Urbanization*, Arturo Soria y Mata, ed. (Barcelona: Electa, 1999)

6 Ildefonso Cerdá, *Teoría de la construcción de las ciudades aplicada al proyecto de reforma y ensanche de Barcelona y otras* conexos (Madrid: Instituto Nacional de la Administración Pública and Ayuntamiento de Madrid, 1859) § 1500

"...such a shift in the relations between body, power, and space is evident in a new mode of urbanism: so-called 'resilient urbanism.'"

To call for such a history is, of course, an indictment of the present—an attempt to illuminate the preconditions of our moment that allow us to anticipate how the codes of the human body may once again be shifting. For my part, I believe that such a shift in the relations between body, power, and space is evident in a new mode of urbanism: so-called "resilient urbanism." Resilient urbanism is essentially smart city urbanism having come of age in the era of climate crisis. If the smart city's techniques aimed to optimize the city, resilient urbanism would adopt these techniques to manage crisis, conceiving the city and its surrounding environment as a single expansive space of data to be monitored and intervened upon in real time.[7] Yet what makes resilient urbanism unique may have less to do with the technologies it deploys than in the cumulative effect they have on the bodies they organize: unlike its twentieth century counterparts, resilient urbanism situates one of its innovations in making-infrastructural the body. The body, in other words, is now a primary site of urbanization.

Resilient Bodies

Though impossible to locate precisely, the origins of resilient urbanism dwell amidst the Cold War cult of individualism in America and the simultaneous emergence of the environmental movement. In this space, a motley assortment of voices—from the bourgeois liberalism of a Jane Jacobs to the military-funded technopositivism of second-order cybernetics—coherent only in their rejection of modernist planning, would unwittingly set out the axes on which a radically new approach to urbanism could be plotted. Resilient urbanism today exists as a network of interests connecting an array of non-governmental initiatives with a host of new courses and degrees offered in top schools of design all aligned with a new attitude in governmental

---

7   Some of the most insightful critical analyses of smart city urbanism can be found in Jennifer Gabrys, "Programming Environments: environmentality and citizen sensing in the smart city," *Environment and Planning D: Society and Space* 32:1 (2014). For an excellent look at its relation to cybernetic discourse, see Maros Krivy, "Toward a critique of cybernetic urbanism: The smart city and the society of control," *Planning Theory* (April, 2016) 1-23.

policymaking. It is perhaps captured most clearly in the Rockefeller-fueled mega-project, Rebuild by Design (RBD), currently under way in the greater NYC region.

On its surface, RBD seems to respond to a certain tendency today of localist, community-oriented, DIY design. It appears very "bottom-up," open-source and, in general, quite happy. It embraces the existing spaces and practices of everyday life in NYC, while conceiving of its interventions as protection against extreme weather. There are in total ten different projects that stretch a collective site along the greater NYC coastal region, all of which embrace strategies that in one way or another construct new relations between urban life and water. Opening urban design as a collaborative effort between architects, marine ecologists, climate scientists, and, tellingly, insurance experts, design becomes less a question of transforming space than of augmenting it—giving it over to new uses, exposing coastal areas to new activities, finding hidden opportunities for a "hopeful" eco-urban life to take root in the spaces that Sandy's destruction inadvertently revealed.[8] Further, there is a clear agenda to rewrite the human relation to nature as one of entanglement. Infrastructure, in this new conception of design—such as systems of flood mitigation and storm surge abatement—is to be designed with and inclusive of natural processes: no longer drawing a boundary separating society from nature, infrastructure now appears as the thing that brings the two together. In its most pronounced examples, infrastructure and nature become indistinguishable from one another in so-called "nature-based solutions."[9]

Yet RBD is also a highly technological space. In as much as it blurs the boundary between infrastructure and nature, it also blurs the line between environment and technology. Indeed, resilient urbanism may be understood as the smart city retooled to mitigate the effects of

---

8   See Orit Halpern, "Hopeful Resilience," *e-flux architecture* (April 19 2017): http://www.e-flux.com/architecture/accumulation/96421/hopeful-resilience/

9   For more on the implications of nature-based solutions in the context of resilient urbanism, see my piece Ross Exo Adams, "An Ecology of Bodies," *Climates: Architecture and the Planetary Imaginary* (New York City/Zürich: Columbia Books on the Architecture and the City/Lars Müller Publishers 2016) 181-190.

climate crisis. In this sense, it expands the application of ubiquitous sensing to include the monitoring of and communication between natural ecologies of the NYC region. Nature-based infrastructures, much like their traditional urban counterparts, are now to be laced with networks of sensors and ubiquitous computing. Furthermore, RBD aims to expand its monitoring capacity by encouraging a culture of "ecological stewardship" to take root in the coastal communities throughout the New York City region. In total, the broad application of environmental sensing is an effort to transform an entire coastal region into a data-intensive and extensive site. As a mode of urbanism, its use of these infrastructures aims to integrate ecosystems and weather patterns through vast new algorithmic techniques of analysis and intervention.

While the full aim of making the NYC region "smart" is never stated directly, a number of proposals make clear that the combination of smart technologies with the eco-cybernetic, "nature-based solutions," constitute a system of crisis management geared to guide bodies that inhabit the region in the event of crisis. Nature, bodies, and infrastructure are to be monitored continuously (what one team calls its "situation analysis") in a space defined by its propensity to crisis. This is why almost all projects go to great lengths to integrate a lexicon and history of environmental crisis into the banality of everyday urban life, through placards marking out flood levels, apps, and displays that purport to monitor levels of risk in real time. Here, "stewardship" reveals another side: its other aim is to provide a vast new trove of data to be farmed at the interface of human life and ecology. When nature-based solutions are seen as data intensive infrastructures, their local practices of care double as techniques to expand the quantity of data that can be mined to correlate the uncertainties of human life and extreme weather as a means to manage both.[10]

---

10   See Rebuild by Design, "Policy by Design: Promoting Resilience in Policy and Practice" (June 2014), http://www.rebuildbydesign.org/data/files/476.pdf

Becoming-infrastructural

Is it plausible to think that resilient urbanism could mark not only the emergence of a new body, but also assist in fundamentally altering the relations between power and space that traverse it? If the rise of the urban in the nineteenth century was a telling response to a new conception of the body, what sort of body might resilient urbanism reflect?

In order for the modern, deficient, exposed, and machinic body to be uniformly governed in the emergent horizon of biopower, space had to be universally transformed into a functional instrument. Yet for early urbanists like Cerdá or Haussmann, this was never given by an overt mandate: space, in other words, was not to be remade as an outcome and representation of some new power structure; rather, the construction of a new spatial order they contributed to would assist in providing a template on which a new, non-representational power-in-space could discover its techniques. For this reason, urbanization for Cerdá, Le Corbusier and many others, remained an ideal project—a "historical duty" of "mankind"—driven by an imperial urge to reify nature and modernize the world. This is perhaps most explicit in Cerdá's work, in which *urbanización* is at once construed as the prehistoric root of humanity *and* its inevitable future.[11] The teleological temporality in which the *urbe* is thought lends itself as a concrete program to "progress," captured by the fact that the urban is also a process whose endpoint Cerdá eagerly imagined as a single urban space stretching across the planet. The body, both drawn into and dependent upon this space, as Cerdá's work attests, is a bio-economic *dividual*,[12] whose multiple divisions dutifully reflect those that organize the *urbe*: both body and *urbe* are joined in their reduction to the binary functions of a capitalist world order—circulation and dwelling, economic accumulation and biological recuperation, waged consumption of labor power and its unwaged reproduction.[13]

---

11   Ildefonso Cerdá, *Teoría general de la urbanización* (Madrid: Imprenta Española, 1867). See also Ross Exo Adams, *Circulation and Urbanization* (London: Sage, in print).

12   Gilles Deleuze, "Postscript on the Societies of Control," *October* 59 (1992) 3-7.

13   Chapter 1 of Ross Exo Adams, Circulation and Urbanization, (London: Sage, in print)

This divided body finds its moral compass in the universal history that motivates the nineteenth century's economic sociology and its desire to reconstruct the world as an 'apolitical' space of unlimited, planetary circulation. Yet far from apolitical, the body had become at once a universal measure of the urban and the object of its biopolitical techniques of normalization and control. At first this happened implicitly, as the nineteenth century architectures and infrastructures that were meant to immunize bodies from themselves, from nature, from disease and insurrection. Then, in the twentieth century, it became explicit in the work of Neufert, Neurath, Corbusier, Dreyfuss and others, in which the body became both a norm used to construct urban space and the irregular objects that the urban aimed to systematically correct.

Resilient urbanism clearly marks a departure from this history. Aesthetically, it rejects any of the aseptic spaces of twentieth century urbanism, or even the idea that space is being transformed, and instead embraces a non-modern attitude that affords romantic, nostalgic relations to a kind of found-object urban space. Its understanding of nature as edgeless and entangled—both process and resource—widens its object of design from the built environment to simply the environment. In this way, bodies, ecologies, and infrastructures become the vectors of a "natural," distributed agency, suggesting a mode of governance no longer seen as external to life, but rather built into a participatory form of self-governance internal to the contours of social and natural complexity.[14]

But these observations only cover over a deeper shift. Indeed, what fundamentally marks a departure from the history of modern urbanism is the way in which resilient urbanism reimagines the body. The environment can only be taken the site of intervention by, at the same time, suggesting a body ontologically internal to it, marked by its malleability and responsiveness to its environment across many scales. Its entangled, more-than-human relation to its environment

---

14 David Chandler, "Beyond neoliberalism; resilience, the new art of governing complexity," *Resilience: International Policies, Practices and Discourses*, 2:1 (2014): 47-63.

thus opens the body up as a site of urban design. The expanded use of ubiquitous sensing technologies, as sociologist Jennifer Gabrys has written, unwittingly turns bodies of resilient urbanism into sensors operating within its cybernetic form of knowledge and algorithmic modes of control.[15] No longer simply the subject of urban design, the body now doubles as its object—as infrastructure—making everyday life indistinguishable from its permanent technological modulation. Resilient urbanism, we could say, is the urbanization of the body.

Post-Historical Bodies

If modern urbanism implemented strategies that sought to eliminate the possibility of crisis, resilient urbanism, we can say, is a project that integrates crisis in *its tactics*; crisis is not something to exclude, but its very condition of possibility. Resilient urbanism is an urbanism of crisis management. The blurring of bodies, natures, and infrastructures reveals a power-in-space built not on standards, norms, or the rule of law, but as a means to engage crisis as its "reality," a condition whose contours can be endlessly extracted from the incomprehensible quantities of data that now constitute the knowledge of the urban—a knowledge of effects without causes.[16] By conceiving the body as a site of urbanization in a space made visible as an arena of crisis, resilient urbanism coincides with the birth of a *post-historical* body—a steward of complexity bound up in the machinic feedback of a nameless, invisible algorithmic governmentality.[17]

In post-history, Vilém Flusser writes, "the present is the totality of the real."[18] If resilient urbanism corresponds to the emergence of a

---

15   See chapters in section III of Jennifer Gabrys, *Program Earth: Environmental Sensing Technology and the Making of a Computational Planet* (Minneapolis: University of Minnesota Press, 2016) 183-265.

16   Antoinette Rouvroy, "The end(s) of critique: data-behaviourism vs. due process" in *Privacy, Due Process and the Computational Turn: The philosophy of law meets the philosophy of technology,* Mireille Hildebrandt and Katja De Vries (eds.) (London: Routledge, 2013)

17   Antoinette Rouvroy, "Algorithmic Governmentality and the End(s) of Critique," lecture, Society of the Query #2, Institute of Network Cultures, Hogeschool van Amsterdam (8 November 2013) https://vimeo.com/79880601

18   Vilém Flusser, *Post-History* trans. Rodrigo Maltez Novaes (Minnesota: Univocal, 2013) 119.

new form of governance, its effects cannot be due simply to its use of environmental sensing or its large-scale rollout of ICT infrastructures. What matters is rather how these technologies make legible a historical, social and political sensibility toward climate change, technology, nature, life, and politics, thus producing it *as reality*. The body of post-history is not a catalyst for this, but serves as both its representation and the diagram for its reproduction. This body, made visible in its eco-cybernetic urbanization, is no longer a *site* of infrastructural control, but infrastructure itself—a shift which profoundly inscribes crisis into the experience of everyday life. In the space of post-history, crisis is the metric of time, experienced as statistical thresholds of the unprecedented and displays of its endless unfolding for all to bear witness. Our exposure to post-historical time, as well as our bodies becoming-infrastructural, de-historicizes the human condition just as it depoliticizes climate change, presenting it as an inevitability for which a new, spectacular optics must be designed. In this space, the modern urgency to accelerate toward a universal, predestined future gives way to a static anxiety of an endless and totalizing present in which "stewardship" substitutes for political agency. Here, tendencies and speculation replace history and futurity and a new diaphanous body-of-effects emerges, coherent only as its digital shadow, registered by its flickering illumination in data space. Yet the experience of an endless present is also the endless production of future scenarios.[19] The body of post-history, transparent and entangled, is thus also a permanent tourist of its own speculation, seduced by images that trade this for a future world ironically animated by fleshy, vibrant bodies brimming with love, intimacy, agency, and consequence—traits which may also reveal a new, somatic horizon of our coming political resistance.

---

19   Orit Halpern, Jesse LeCavalier, Nerea Calvillo and Wolfgang Pietsch, "Test-Bed Urbanism," *Public Culture*, 25:2 (2013) 272-306.

Breaking
the Waves

By Olivia Rosane

(First appeared in *Real Life*,
November 2017)

I spent this summer wandering along the high-water mark of future floods. As part of a project to envision a climate-changed London, I walked through Thames-front neighborhoods, trying to determine which parks and buildings would be oversurged by storms and tides and which would rise high enough for survivors to use as shelter. My guide in this effort was a map, "Surging Seas, Mapping Choices," designed in the run-up to the 2015 UN Climate Change Conference that resulted in the Paris Agreement (the accord that President Trump recently pulled the U.S. out of). The map — made by Stamen Design, a San Francisco firm, using research conducted by Climate Central, a climate reporting organization run by journalists and scientists — allows viewers to compare the water levels projected to result from various global warming scenarios. Just type the name of a coastal location into the search bar, and the map shows you water levels after two degrees Celsius of warming (a long-discussed international target) and four degrees Celsius (the predicted consequence of business as usual).

As its name implies, the map is intended as a warning and a call to action. Overlaid fonts in traffic-cone orange and yield-sign yellow try to visually mark the urgency. But on the map itself the still dry land is coded a calming slate-gray, while the rising water is a soothing middle blue. Zoom in close enough, and you can see transparent aquamarine washing over the satellite images of buildings, the new shoreline always composed of crisp, rational straight lines.

The map is too precise and attractive to instill fear. Instead, it makes an out-of-control future look exactly predictable: These buildings will be swallowed, but not those. And the swallowing appears as such a clean process. The buildings beneath the waterline remain intact, merely covered with blue. There's no suggestion of damage or debris. And the difference between water levels is the work of a mouse-click.

The designers of the map hoped to encourage viewers to believe that they could determine the future by lobbying governments to push for lower emissions in Paris, but the map itself suggests that control over the climate could be much more direct and much less hard-fought. Don't worry, the graphics say, even if the water rises, it will rise in a neat and orderly progression.

Wandering through London with this map in my palm, down streets without vistas bordered by rows of brick houses, I found it hard to imagine what the world it predicted would look like, unpixelated and in three dimensions. The district of Wapping, for example, whose Rose Garden will be an island after four degrees of warming, is a place of narrow cobblestone streets and quaint, light-blocking row homes. You can't always see the river even when a map tells you you're walking beside it. I found a gap between a brick pub and a private dock, and it was like a secret door to a different world: I descended a flight of seaweed-coated steps and stood on a beach of pebbles brightened with terra-cotta pottery shards, looking out at a gray river lapping at the stones with a sound like sleep-deep breathing.

The tranquility of the scene matched the calming lines and colors on the Surging Seas map. Looking from Thames to palm, it was possible to gently imagine this space expanded up to the green grass of the Rose Garden, the red brick and pale cobbles of the surrounding streets long since broken into pebbles and lapped by waves on a new beach. Neither image helped me visualize the violent transition that would occur between the two, the winds and waves that would break down the buildings and scatter the people inside them. There is no connection between the clear blue and crisp lines of the Climate Central map and the recent photographs of flooding I've seen coming out of Houston, Bangladesh, and Puerto Rico, of high-piled debris and streets turned into muddy rivers.

The Climate Central map was an invaluable resource, and I relied on it so heavily for my project that it feels ungrateful to reprimand it. And it is by no means unique in having an aesthetically pleasing presentation that runs counter to the devastating consequences it seeks to represent. *National Geographic*, for example, published a series of maps inviting readers to "explore what the world's new coastlines would look like" if all the sea ice melted, depicting the continents in resplendent, Earth-from-space jewel tones. Geology.com's sea-level-rise map has a relatively primitive, early-web design, but it still allows you to zoom in on different locations and alter sea-level rise with a drop-down arrow key, as if you were playing a very basic computer game. Its colors are also pleasing cobalt blues for sea and pale greens or light ochres for land and not the turgid browns and grays of actual storm surges.

Temperature projections also can be surprisingly attractive. A map showing that the average summer temperature in North Africa and the Middle East will increase by five degrees Celsius by mid-century if business continues as usual paints the region in a vibrant crimson offset by patches of hotter maroon and cooler tangerine. The image is dotted with small black circles, which actually indicate that all 26 climate models used to predict the increase are in agreement, but aesthetically they still soften the visual impact of the map. The bright colors and regular pattern suggest the fabric for a child's summer dress.

Designed to show inputs and outputs, maps and models can't easily represent process. They tend to obscure the truth that the main danger of climate change is *change*: which means mess and violence and fleshy bodies against heat and water. The dominance of such maps in the visual culture of climate-change discourse makes the process of change appear much neater and more controllable than it actually has been or will be.

There is a delicate balance between depicting climate change as urgent enough of a problem to inspire action, and inundating viewers with paralyzing images of apocalypse. In "Communicating Climate Change: Closing the Science-Action Gap," Susanne C. Moser and Lisa Dilling discuss how fear-based appeals without any accompanying action plan lead to "denial, numbing, and apathy." The designers of these maps do not want to frighten people away from engaging. However, the false reassurance of the aesthetically pleasing images can lead to a different kind of paralysis: the false political reassurance that the panels of experts, private interests, and global summits that have been consistently convened to try to get

"There is a delicate balance between depicting climate change as urgent enough of a problem to inspire action, and inundating viewers with paralyzing images of apocalypse."

us out of this crisis will eventually get it right without our having to do anything.

Take Mission 2020, which sets itself the laudable goal of "bending the curve" on carbon emissions by 2020, with an eye on reaching zero emissions by 2050. On its website one sees first a statement of purpose in bold white against a gradient of crimson and orange. The background choice is confusing, since it at once suggests the hotter temperatures the Mission hopes to avoid and makes them look surprisingly appealing. The white text laying out the goal is especially striking against the bright orange background; it does not look like a future to avoid at all costs. Scrolling down, one next sees a lineup of six hoped-for milestones in bright, primary colors: for example, "renewables outperform fossil fuels" and "heavy industry ... commits to being Paris compliant." There is apparently nothing the ordinary visitor needs to do to bring about this bright future but scroll down even further and buy a T-shirt that reads "Stubborn Climate Optimist."

Mission 2020's chances of success appear to rely entirely on the innovations of the well-educated and the actions of the powerful. Its website proclaims its benchmark to be "achievable" because "renewables are rapidly falling in price," "technological progress on battery storage" is improving the performance of renewables and electric cars, and business leaders, cities, and the financial community are working on solutions. There is no mention of the fact that chasing cheaper, more efficient energy is how we warmed the planet in the first place. Nor does the website address the question of what happens when profits and climate action clash. The auto industry might be coming around on electric cars now, but in 2001 GM famously killed their own electric model after lobbying successfully to weaken the California legislation that had required they build it. And if renewables are doing well now, it is partly thanks to the activists who pushed for laws requiring states and countries get a certain percentage of their energy from clean sources; half the investments in Germany's green energy boom came from individual citizens or citizens' energy associations.

The mission is spearheaded by Christiana Figueres, the former executive secretary of the UN Framework Convention on Climate Change who helped broker the 2015 Paris Agreement, and I sincerely wish her every success in this endeavor. But it seems naive to expect business leaders, governments, and financial tycoons to turn the tide on climate change without massive grassroots pressure from the rest of us. Now is not the time to be lulled into the dream that we can sit back and watch the corporate PowerPoint presentation. Yet this is the aesthetic the Mission 2020 site most suggests.

Why do mainstream climate organizations rely on the visual language of data, focusing on representing "degrees hotter" and "meters higher" instead of lives lost or homes swallowed? By placing an emphasis on the facts of climate change rather than its human consequences, centrist liberals can turn climate denial into a bigger enemy than climate inaction. In a recent segment, for example, Samantha Bee led a group of climate skeptics through a "Hell House" of climate-change predictions. The punch line was that the one guest to change her mind changed it not from fear of any vision in the house but because the other invitees were so obviously ignorant, she feared to be publicly associated with their views. By focusing on changing minds, with a few jokes at the expense of the dummies thrown in, you never have to answer the truly frightening question of what to do once all the minds have been changed.

The centrist use of fact dissemination as a bulwark against radical action extends beyond climate change. Writing after Brexit and the election of Donald Trump, Bue Rübner Hansen and Rune Møller Stahl argue that liberals have emphasized a vision of government as fact-based administration. In fact, one of the great lies of the neoliberal era was the assurance that formerly political concerns could now be managed by experts rather than debated by real life stakeholders. In Francis Fukuyama's "The End of History," one of that era's opening salvos, he claimed wistfully that, after the fall of the Berlin Wall, the "worldwide ideological struggle" would be replaced by "the endless solving of technical problems" and "environmental concerns."

One example of this mentality in action is the graphic and map-filled election night coverage, which, up until 2017's surprise, might have tricked viewers into thinking the results have more to do with

Nate Silver's projections than with door knocking. According to *Politico*, the Clinton campaign even ordered a group of SEIU members, who were anxious about Michigan and wanted to canvas there, to turn around and return to Iowa because the campaign's models told them Michigan was safe.

Brexit and Trump prove definitively that the expert control of politics is over. But climate change still presents a special dilemma for anyone clinging to both fact-based leadership and neoliberal capitalism. All the smartest people in the relevant fields agree that climate change is happening and is caused by humans. And all those people haven't been able to do anything about it because capitalism has gotten in the way. The first Intergovernmental Panel on Climate Change was established in 1988, a year before Fukuyama penned his essay. Five IPCCs later, and 2016 was the hottest year on record.

That is because climate change is not merely a technical problem; it is an inherently political one: a question of who has the power to profit from polluting and who is left to pay the price, in flooded homes, droughts, and worse. This is not to say that technology can't help curb emissions, but its effectiveness will always be constrained by political forces. For instance, Naomi Klein, in an interview about her book *This Changes Everything*, discusses how Germany has managed to get 25 percent of its energy from renewable resources yet its emissions are still up because the country still mines and exports coal, and "Angela Merkel refuses to stand up to the coal lobby." World Trade Organization rules often hamper green energy innovation. In another example from Klein's book, a law designed to wean Ontario off coal by 2014 was struck down by the WTO because it allowed renewable energy companies to sell power back to the grid if 40 to 60 percent of their materials and workers were local, something that violated the organization's ban on laws that distinguish between local and international companies.

Climate change will not be solved by appealing to the status quo because the status quo is designed to protect the profits of international firms above all else. No wonder then that centrist liberals, forced to choose between markets and facts, retreat into graphic fantasy. Stamen Design, the firm behind the Climate Central map, also did data visualization for Toyota i-Road, a sustainable, three-wheeled vehicle, ahead of its 2015 test drive by tech luminaries and car enthusiasts in San Francisco. When corporate greenwashing and climate-change models share a glossy visual language, all sense of struggle, in both the hopeful and harrowing sense of the word, is erased. Climate change then appears like something safely out of ordinary hands, when in fact it will take all hands to slow it and to care for those it has already displaced.

Scientist Brad Werner developed a computer model of interactions between humans and the environment that revealed that the one human action that had a chance against climate change was popular resistance on levels analogous to the civil rights or anti-slavery movements. According to the summary of his research at *Slate*, "every other element — environmental regulation, even science—is too embedded in the dominant economic system." What's needed is to jolt nonscientists, ordinary people, out of the illusion that they can rely on staid experts to resolve this crisis. In a small effort toward this end, Werner gave his presentation of this research at the 2012 American Geological Union meeting a pointed title: "Is Earth F**cked?" He said the irreverence of his title was sparked by his friends who were depressed not by the good climate science being done to predict the future "but by the seeming inability to respond appropriately to it."

That doesn't mean there isn't an appropriate response. It's a response that refuses to wait on politicians or technocrats. Klein refers to a "Blockadia" movement, a loose network of ordinary people putting their bodies in between the earth and new means of extracting fossil fuels, as at the protest camp at Standing Rock against the Dakota Access Pipeline. But this too is dangerous and messy, much harder than choosing a certain number of degrees of warming on an interactive map. When you look at the videos that emerged from the camp resisting the Dakota Access Pipeline, you see an eerie echo of the news images of floods and hurricanes: fleshy bodies against water.

It may be that recognizing ourselves as this vulnerable is the only way we will save ourselves. Right now, the communities leading the direct action fight against climate change are often, like the Standing Rock Sioux, indigenous groups who are fighting the encroachment of fossil fuel companies on their local environments. As the already vulnerable will disproportionately suffer from climate change itself, the already vulnerable pay a disproportionate price to fight it: 40 percent of those murdered for environmental activism in 2015 were from indigenous communities. For many of these groups, fighting climate change is not an abstract mission but part of the work of caring for and protecting their very specific homes.

It is that local detail that is partly missing in maps and models intended to synthesize data points into a single narrative. The neat images that summarize that work obscure the labor that went into them as well as the action needed to prevent them. In an interview in *Art in the Anthropocene*, Bruno Latour talks about how Charles Keeling, whose measurements of carbon dioxide in the atmosphere at the Mauna Loa Observatory in Hawaii first raised the possibility of anthropogenic climate change, got his data by "being there for everyday for 30 years, or even 50 years." He calls this patient observation, "the science of care." That sense of care is lost when you look at the zipper-like upsweep of the Keeling Curve.

But it is care that must be central to the discussion around climate change, because it is care that has been sidelined by the economic forces that unleashed it. Lowering emissions in time to avoid more hurricanes means committing to the belief that nothing — no invention, no profit motive, no growth opportunity — is more important than the preservation and flourishing of life, human and otherwise. Perhaps such care cannot be effectively mapped. It can only be enacted.

Why Hope Is Dangerous When It Comes to Climate Change

By Tommy Lynch

(Originally appeared in *Slate*, July 2017)

L ots of people worry about climate change, but as David Wallace-Wells shows in his recent *New York* magazine piece, the future is almost certainly worse than you imagine. Drawing on a wide range of experts, he tracks how climate change could alter every aspect of planetary existence. Ocean acidification gives rise to oxygen-eating bacteria. Melting ice results in the absorption of more sunlight and greater warming. Rising temperatures hasten the destruction of plants that replenish our oxygen. As things get worse, they will get worse faster.

Given the thoroughness of Wallace-Wells' evidence, the ending comes as a bit of a surprise.

> We have not developed much of a religion of meaning around climate change that might comfort us, or give us purpose, in the face of possible annihilation. But climate scientists have a strange kind of faith: We will find a way to forestall radical warming, they say, because we must.

The same "strange kind of faith" is behind condemnations of the piece as alarmist. Some climate scientists have questioned Wallace-Wells' treatment of the evidence. Radical warming can be slowed, they say, but if journalists or scientists scare people they risk disrupting the important work that needs to be done. The climate scientist Michael Mann, in a widely circulated Facebook post, worries about the "danger in overstating the science in a way that presents the problem as unsolvable, and feeds a sense of doom, inevitability and hopelessness." The fear is that people won't devote the necessary political and economic resources to these problems if there isn't some hope that it will work out in the end.

When we look at more mainstream predictions, however, there doesn't seem to be much reason for hope. Although we are unlikely to experience the "doomsday" scenario described by Wallace-Wells, we will likely see increases that will exacerbate existing inequalities as we experience changes in weather patterns that affect life in coastal cities, the production of food, and global conflicts (as Mann himself explains). Even if things aren't going to be as bad as the worst-case scenario, the future still isn't looking good.

As concerns about climate change have intensified, philosophers have increasingly devoted attention to how we might balance hopefulness with confronting the ways that the climate is already changing. Like the scientists who spoke to Wallace-Wells, many philosophers worry that pessimism is a threat to this work. For example, ethicist Kathryn Norlock has written on the importance of maintaining hope even when pessimism is a rational response. The burden of hope falls particularly on those who live in affluent societies. Indulging despair would risk sabotaging any adequate collective response to the situation. We should also resist the temptation to single out groups of people as responsible for climate change. Instead, we should forgive those we think are guilty of environmental harm in order to maximize our ability to work together for a better world. Now is not the time for blame, Norlock says, but for new forms of ecocitizenship.

Hope that science will provide a solution is its own kind of surrender.

Norlock's argument makes sense on one level. Relatively affluent people are free to throw up their hands in defeat at the prospect of climate change, safe in the knowledge that they (and their children) have the resources to mitigate its consequences, at least for a little while. If there's nothing to be done, you might as well enjoy things while you can. It is important to combat this resignation, but resignation and hope aren't our only options. Though there are risks to embracing pessimism and fear, they are a necessary aspect of confronting our situation. And more positive outlooks entail their own problems. Hoping that science will provide a solution is its own kind of surrender, relieving the pressure of confronting the ways of life that have given rise to climate change in the first place. This hope also downplays the fact that such solutions likely will entail living in a world marked by pain and suffering directly and indirectly caused by what we have done to nature.

These demands that we hope against all evidence are examples of what Lauren Berlant calls "cruel optimism." Berlant describes the way people hope for something that is impossible or fantastical. What makes this cruel, rather than just tragic, is that the hope is itself part of the problem. Think of the way that dreams of success and wealth function in American society. Low-paid employees in precarious positions are told that determination and hard work will result in greater opportunities and economic security. In actuality, class mobility is very limited. The optimism at the heart of the American dream is cruel: Workers invest in a dream that actually leaves them more open to exploitation rather than challenging the wider economic system.

Berlant's "cruel optimism" is a useful way of thinking about the demand to stay hopeful in the face of climate change. The hope that we will invent technological means of preserving our way of life is itself part of the problem. It is not that we live in a world where our economics, politics and culture happen to contribute to climate change, but that life in "the West" is essentially destructive of the rest of nature. As sociologist Jason Moore explains, we depend on "cheap nature"—the stores of energy and raw materials that we extract from the earth. Climate change results from activities that are rapidly depleting those stores and the consequences of climate change mean the stores won't be replenished anytime soon. The problem isn't an accidental byproduct of our way of life—it's our very way of life.

The term *Anthropocene*, once confined to academic journals and conferences, is now casually dropped in podcasts and splashed across magazines like *Slate*. It refers to the geological epoch in which humanity became a force that changed the environment. Moore suggests using *Capitalocene* as an alternative to *Anthropocene*. As the *Guardian* reports, a recent study shows that 100 companies are responsible for 71 percent of the greenhouse-gas emissions since 1998. While almost all people play some role in the degradation of the environment, climate change is also something done to people by other people. It isn't humanity as such that is responsible, but the specific forms

"Obviously not all people experience this world in the same way, and it is a further tragedy that those who have contributed the least to climate change will be among those who experience its consequences earliest."

of production and consumption that are the basis of the capitalist Western world.

That world is ending: a world of eating food shipped from country to country, a world of discount airlines, widespread meat consumption, and constant air conditioning. The problem with hoping for a technological solution to climate change is that it is often insufficiently critical of the ways of life that wreaked havoc on the rest of nature. It is easier to hope for a wild geoengineering solution than face the reality that billions of people need to change their daily habits in order to lessen the immense suffering appearing on the horizon. This hope cruelly prevents us from confronting the deep structural challenge of rethinking the way that some humans relate to nature. Obviously not all people experience this world in the same way, and it is a further tragedy that those who have contributed the least to climate change will be among those who experience its consequences earliest.

Some responses to Wallace-Wells' piece have decried its alarmism and despair. But *Slate*'s Susan Matthews has already argued that it is not alarmist enough. I agree—and I would add that its hopeful conclusion also avoids the pessimism necessary for confronting the reality of the changes ahead.

Pessimism isn't popular at the moment. As Jill Lepore wrote in the *New Yorker* earlier this summer, "Radical pessimism is a dismal trend." Considering recent novels that offer pessimistic pictures of political and ecological futures, she concludes that dystopia is no longer "a fiction of resistance." It despairs instead of calling for action.

The accusation that pessimism results in political paralysis is frequently made in the process of advocating hope. In the most recent edition of her book *Hope in the Dark*, Rebecca Solnit argues that pessimists focus on disappointment as a way of avoiding taking action, decrying every possibility as imperfect and inadequate. She differentiates hope from both optimism and pessimism by its acceptance of uncertainty. Optimists think everything will be fine, pessimists think everything will be terrible, but those who are hopeful act in the belief that actions will, in some way and at some point, matter.

Solnit's uncertain hope, while not naive optimism, still does not help us answer the fundamental questions posed by climate change:

What should we hope for? What *shouldn't* we hope for? What should we hope against? Solnit, like many, poses pessimism and hope as two mutually exclusive options. Yet the first can be a condition for the second. We cannot answer the question "What should we hope for?" without confronting that for which we should despair.

If Moore is right, then the patterns of production and consumption at the heart of the global economy are integral to global warming. Maybe that way of life isn't worth saving. Kafka reportedly once said that there is "plenty of hope, an infinite amount of hope—but not for us." Rather than investing in technological salvations that will allow us to prolong a way of life that is destroying the rest of nature, we can embrace pessimism. In abandoning hope that one way of life will continue, we open up a space for alternative hopes.

Estonia, the Digital Republic

By Nathan Heller

(Originally appeared in
the *New Yorker*,
December 2017)

U p the Estonian coast, a five-lane highway bends with the path of the sea, then breaks inland, leaving cars to follow a thin road toward the houses at the water's edge. There is a gated community here, but it is not the usual kind. The gate is low—a picket fence—as if to prevent the dunes from riding up into the street. The entrance is blocked by a railroad-crossing arm, not so much to keep out strangers as to make sure they come with intent. Beyond the gate, there is a schoolhouse, and a few homes line a narrow drive. From Tallinn, Estonia's capital, you arrive dazed: trees trace the highway, and the cars go fast, as if to get in front of something that no one can see.

Within this gated community lives a man, his family, and one vision of the future. Taavi Kotka, who spent four years as Estonia's chief information officer, is one of the leading public faces of a project known as e-Estonia: a coördinated governmental effort to transform the country from a state into a digital society.

E-Estonia is the most ambitious project in technological statecraft today, for it includes all members of the government, and alters citizens' daily lives. The normal services that government is involved with—legislation, voting, education, justice, health care, banking, taxes, policing, and so on—have been digitally linked across one platform, wiring up the nation. A lawn outside Kotka's large house was being trimmed by a small robot, wheeling itself forward and nibbling the grass.

"Everything here is robots," Kotka said. "Robots here, robots there." He sometimes felt that the lawnmower had a soul. "At parties, it gets *close* to people," he explained.

A curious wind was sucking in a thick fog from the water, and Kotka led me inside. His study was cluttered, with a long table bearing a chessboard and a bowl of foil-wrapped wafer chocolates (a mark of hospitality at Estonian meetings). A four-masted model ship was perched near the window; in the corner was a pile of robot toys.

"We had to set a goal that resonates, large enough for the society to believe in," Kotka went on.

He is tall with thin blond hair that, kept shaggy, almost conceals its recession. He has the liberated confidence, tinged with irony, of a cardplayer who has won a lot of hands and can afford to lose some chips.

It was during Kotka's tenure that the e-Estonian goal reached its fruition. Today, citizens can vote from their laptops and challenge parking tickets from home. They do so through the "once only" policy, which dictates that no single piece of information should be entered twice. Instead of having to "prepare" a loan application, applicants have their data—income, debt, savings—pulled from elsewhere in the system. There's nothing to fill out in doctors' waiting rooms, because physicians can access their patients' medical histories. Estonia's system is keyed to a chip-I.D. card that reduces typically onerous, integrative processes—such as doing taxes—to quick work. "If a couple in love would like to marry, they still have to visit the government location and express their will," Andrus Kaarelson, a director at the Estonian Information Systems Authority, says. But, apart from transfers of physical property, such as buying a house, all bureaucratic processes can be done online.

Estonia is a Baltic country of 1.3 million people and four million hectares, half of which is forest. Its government presents this digitization as a cost-saving efficiency and an equalizing force. Digitizing processes reportedly saves the state two per cent of its G.D.P. a year in salaries and expenses. Since that's the same amount it pays to meet the *NATO* threshold for protection (Estonia—which has a notably vexed relationship with Russia—has a comparatively small military), its former President Toomas Hendrik Ilves liked to joke that the country got its national security for free.

Other benefits have followed. "If everything is digital, and location-independent, you can run a borderless country," Kotka said. In 2014, the government launched a digital "residency" program, which allows logged-in foreigners to partake of some Estonian services, such as banking, as if they were living in the country. Other measures encourage international startups to put down virtual roots; Estonia has the lowest business-tax rates in the European Union, and has become known for liberal regulations around tech research. It is legal to test Level 3 driverless cars (in which a human driver can take control) on all Estonian roads, and the country is planning ahead for Level 5 (cars that take off on their own). "We believe that innovation happens anyway," Viljar Lubi, Estonia's deputy secretary for economic development, says. "If we close ourselves off, the innovation happens somewhere else."

"It makes it so that, if one country is not performing as well as another country, people are going to the one that is performing better—competitive governance is what I'm calling it," Tim Draper, a venture capitalist at the Silicon Valley firm Draper Fisher Jurvetson and one of Estonia's leading tech boosters, says "We're about to go into a very interesting time where a lot of governments can become virtual."

Previously, Estonia's best-known industry was logging, but Skype was built there using mostly local engineers, and countless other startups have sprung from its soil. "It's not an offshore paradise, but you can capitalize a lot of money," Thomas Padovani, a Frenchman who co-founded the digital-ad startup Adcash in Estonia, explains. "And the administration is light, all the way." A light touch does not mean a restricted one, however, and the guiding influence of government is everywhere.

As an engineer, Kotka said, he found the challenge of helping to construct a digital nation too much to resist. "Imagine that it's your task to build the Golden Gate Bridge," he said excitedly. "You have to change the whole way of thinking about society." So far, Estonia is past halfway there.

One afternoon, I met a woman named Anna Piperal at the e-Estonia Showroom. Piperal is the "e-Estonia ambassador"; the showroom is a permanent exhibit on the glories of digitized Estonia, from Skype to Timbeter, an app designed to count big piles of logs. (Its founder told me that she'd struggled to win over the wary titans of Big Log, who preferred to count the inefficient way.) Piperal has blond hair and an air of brisk, Northern European professionalism. She pulled out her I.D. card; slid it into her laptop, which, like the walls of the room, was faced with blond wood; and typed in her secret code, one of two that went with her I.D. The other code issues her digital signature—a seal that, Estonians point out, is much harder to forge than a scribble.

"This *PIN* code just starts the whole decryption process," Piperal explained. "I'll start with my personal data from the population registry." She gestured toward a box on the screen. "It has my document numbers, my phone number, my e-mail account. Then there's real estate, the land registry." Elsewhere, a box included all of her employment information; another contained her traffic records and her car insurance. She pointed at the tax box. "I have no tax debts; otherwise, that would be there. And I'm finishing a master's at the Tallinn University of Technology, so here"—she pointed to the education box—"I have my student information. If I buy a ticket, the system can verify, automatically, that I'm a student." She clicked into the education box, and a detailed view came up, listing her previous degrees.

"My cat is in the pet registry," Piperal said proudly, pointing again. "We are done with the vaccines."

Data aren't centrally held, thus reducing the chance of Equifax-level breaches. Instead, the government's data platform, X-Road, links individual servers through end-to-end encrypted pathways, letting information live locally. Your dentist's practice holds its own data; so does your high school and your bank. When a user requests a piece of information, it is delivered like a boat crossing a canal via locks.

Although X-Road is a government platform, it has become, owing to its ubiquity, the network that many major private firms build on, too. Finland, Estonia's neighbor to the north, recently began using X-Road, which means that certain data—for instance, prescriptions that you're able to pick up at a local pharmacy—can be linked between the nations. It is easy to imagine a novel internationalism taking shape in this form. Toomas Ilves, Estonia's former President and a longtime driver of its digitization efforts, is currently a distinguished visiting fellow at Stanford, and says he was shocked at how retrograde U.S. bureaucracy seems even in the heart of Silicon Valley. "It's like the nineteen-fifties—I had to provide an electrical bill to prove I live here!" he exclaimed. "You can get an iPhone X, but, if you have to register your car, forget it."

X-Road is appealing due to its rigorous filtering: Piperal's teachers can enter her grades, but they can't access her financial history, and even a file that's accessible to medical specialists can be sealed off from other doctors if Piperal doesn't want it seen.

"I'll show you a digital health record," she said, to explain. "A doctor from here"—a file from one clinic—"can see the research that this doctor"—she pointed to another—"does." She'd locked a third record, from a female-medicine practice, so that no other doctor would be able to see it. A tenet of the Estonian system is that an individual owns all information recorded about him or her. Every time a doctor (or a border guard, a police officer, a banker, or a minister) glances at any of Piperal's secure data online, that look is recorded and reported. Peeping at another person's secure data for no reason is a criminal offense. "In Estonia, we don't have Big Brother; we have Little Brother," a local told me. "You can tell him what to do and maybe also beat him up."

Business and land-registry information is considered public, so Piperal used the system to access the profile of an Estonian politician. "Let's see his land registry," she said, pulling up a list of properties. "You can see there are three land plots he has, and this one is located"—she clicked, and a satellite photograph of a sprawling beach house appeared—"on the sea."

The openness is startling. Finding the business interests of the rich and powerful—a hefty field of journalism in the United States—takes a moment's research, because every business connection or investment captured in any record in Estonia becomes searchable public information. (An online tool even lets citizens map webs of connection, follow-the-money style.) Traffic stops are illegal in the absence of a moving violation, because officers acquire records from a license-plate scan. Polling-place intimidation is a non-issue if people can vote—and then change their votes, up to the deadline—at home, online. And heat is taken off immigration because, in a borderless society, a resident need not even have visited Estonia in order to work and pay taxes under its dominion.

Soon after becoming the C.I.O., in 2013, Taavi Kotka was charged with an unlikely project: expanding Estonia's population. The motive was predominantly economic. "Countries are like enterprises," he said. "They want to increase the wealth of their own people."

Tallinn, a harbor city with a population just over four hundred thousand, does not seem to be on a path toward outsized growth. Not far from the cobbled streets of the hilly Old Town is a business center, where boxy Soviet structures have been supplanted by stylish buildings of a Scandinavian cast. Otherwise, the capital seems pleasantly preserved in time. The coastal daylight is bright and thick, and, when a breeze comes off the Baltic, silver-birch leaves shimmer like chimes. "I came home to a great autumn / to a luminous landscape," the Estonian poet Jaan Kaplinski wrote decades ago. This much has not changed.

Kotka, however, thought that it was possible to increase the population just by changing how you thought of what a population was. Consider music, he said. Twenty years ago, you bought a CD and played the album through. Now you listen track by track, on demand. "If countries are competing not only on physical talent moving to their country but also on how to get the best virtual talent *connected* to their country, it becomes a disruption like the one we have seen in the music industry," he said. "And it's basically a zero-cost project, because we already have this infrastructure for our own people."

The program that resulted is called e-residency, and it permits citizens of another country to become residents of Estonia without ever visiting the place. An e-resident has no leg up at the customs desk, but the program allows individuals to tap into Estonia's digital services from afar.

I applied for Estonian e-residency one recent morning at my apartment, and it took about ten minutes. The application cost a hundred euros, and the hardest part was finding a passport photograph to upload, for my card. After approval, I would pick up my credentials in person, like a passport, at the Estonian Consulate in New York.

This physical task proved to be the main stumbling block, Ott Vatter, the deputy director of e-residency, explained, because consulates were reluctant to expand their workload to include a new document. Mild xenophobia made some Estonians at home wary, too. "Inside Estonia, the mentality is kind of 'What is the gain, and where is the money?' " he said. The physical factor still imposes limitations—only thirty-eight consulates have agreed to issue documents, and they are distributed unevenly. (Estonia has only one embassy in all of Africa.) But the office has made special accommodations for several popular locations. Since there's no Estonian consulate in San Francisco, the New York consulate flies personnel to California every three months to batch-process Silicon Valley applicants.

"I had a deal that I did with Funderbeam, in Estonia," Tim Draper, who became Estonia's second e-resident, told me. "We decided to use a 'smart contract'—the first ever in a venture deal!" Smart contracts are encoded on a digital ledger and, notably, don't require an outside administrative authority. It was an appealing prospect, and Draper, with his market investor's gaze, recognized a new market for élite tech brainpower and capital. "I thought, Wow! Governments are going to have to compete with each other for us," he said.

So far, twenty-eight thousand people have applied for e-residency, mostly from neighboring countries: Finland and Russia. But Italy and Ukraine follow, and U.K. applications spiked during Brexit. (Many applicants are footloose entrepreneurs or solo venders who want to be based in the E.U.) Because eighty-eight per cent of applicants are men, the United Nations has begun seeking applications for female entrepreneurs in India.

"There are so many companies in the world for whom working across borders is a big hassle and a source of expense," Siim Sikkut, Estonia's current C.I.O., says. Today, in Estonia, the weekly e-residency application rate exceeds the birth rate. "We tried to make more babies, but it's not that easy," he explained.

With so many businesses abroad, Estonia's startup-ism hardly leaves an urban trace. I went to visit one of the places it does show:

a co-working space, Lift99, in a complex called the Telliskivi Creative City. The Creative City, a former industrial park, is draped with trees and framed by buildings whose peeling exteriors have turned the yellows of a worn-out sponge. There are murals, outdoor sculptures, and bills for coming shows; the space is shaped by communalism and by the spirit of creative unrule. One art work consists of stacked logs labelled with Tallinn startups: Insly, Leap*IN*, Photry, and something called 3D Creationist.

The office manager, Elina Kaarneem, greeted me near the entrance. "Please remove your shoes," she said. Lift99, which houses thirty-two companies and five freelancers, had industrial windows, with a two-floor open-plan workspace. Both levels also included smaller rooms named for techies who had done business with Estonia. There was a Zennström Room, after Niklas Zennström, the Swedish entrepreneur who co-founded Skype, in Tallinn. There was a Horowitz Room, for the venture capitalist Ben Horowitz, who has invested in Estonian tech. There was also a Tchaikovsky Room, because the composer had a summer house in Estonia and once said something nice about the place.

"This is not the usual co-working space, because we choose every human," Ragnar Sass, who founded Lift99, exclaimed in the Hemingway Room. Hemingway, too, once said something about Estonia; a version of his pronouncement—"No well-run yacht basin is complete without at least two Estonians"—had been spray-stencilled on the wall, along with his face.

The room was extremely small, with two cushioned benches facing each other. Sass took one; I took the other. "Many times, a miracle can happen if you put talented people in one room," he said as I tried to keep my knees inside my space. Not far from the Hemingway Room, Barack Obama's face was also on a wall. Obama Rooms are booths for making cell-phone calls, following something he once said about Estonia. ("I should have called the Estonians when we were setting up our health-care Web site.") That had been stencilled on the wall as well.

Some of the companies at Lift99 are local startups, but others are international firms seeking an Estonian foothold. In something called the Draper Room, for Tim Draper, I met an Estonian engineer, Margus Maantoa, who was launching the Tallinn branch of the German motion-control company Trinamic. Maantoa shares the room with other companies, and, to avoid disturbing them, we went to the Iceland Room. (Iceland was the first country to recognize Estonian independence.) The seats around the table in the Iceland Room were swings.

I took a swing, and Maantoa took another. He said, "I studied engineering and physics in Sweden, and then, seven years ago, I moved back to Estonia because so much is going on." He asked whether I wanted to talk with his boss, Michael Randt, at the Trinamic headquarters, in Hamburg, and I said that I did, so he opened his laptop and set up a conference call on Skype. Randt was sitting at a table, peering down at us as if we were a mug of coffee. Tallinn had a great talent pool, he said: "Software companies are absorbing a lot of this labor, but, when it comes to hardware, there are only a few companies around." He was an e-resident, so opening a Tallinn office was fast.

Maantoa took me upstairs, where he had a laboratory space that looked like a janitor's closet. Between a water heater and two large air ducts, he had set up a desk with a 3-D printer and a robotic motion-control platform. I walked him back to Draper and looked up another startup, an Estonian company called Ööd, which makes one-room, two-hundred-square-foot huts that you can order prefab. The rooms have floor-to-ceiling windows of one-way glass, climate control, furniture, and lovely wood floors. They come in a truck and are dropped into the countryside.

"Sometimes you want something small, but you don't want to be in a tent," Kaspar Kägu, the head of Ööd sales, explained. "You want a shower in the morning and your coffee and a beautiful landscape. Fifty-two per cent of Estonia is covered by forestland, and we're rather introverted people, so we want to be—uh, not near everybody else." People of a more sociable disposition could scatter these box homes on their property, he explained, and rent them out on services like Airbnb.

"We like to go to nature—but comfortably," Andreas Tiik, who founded Ööd with his carpenter brother, Jaak, told me. The company had queued preorders from people in Silicon Valley, who also liked the idea, and was tweaking the design for local markets. "We're building a sauna in it," Kägu said.

In the U.S., it is generally assumed that private industry leads innovation. Many ambitious techies I met in Tallinn, though, were leaving industry to go work for the state. "If someone had asked me, three years ago, if I could imagine myself working for the government, I would have said, 'Fuck no,' " Ott Vatter, who had sold his own business, told me. "But I decided that I could go to the U.S. at any point, and work in an average job at a private company. This is so much bigger."

The bigness is partly inherent in the government's appetite for large problems. In Tallinn's courtrooms, judges' benches are fitted with two monitors, for consulting information during the proceedings, and case files are assembled according to the once-only principle. The police make reports directly into the system; forensic specialists at the scene or in the lab do likewise. Lawyers log on—as do judges, prison wardens, plaintiffs, and defendants, each through his or her portal. The Estonian courts used to be notoriously backlogged, but that is no longer the case.

"No one was able to say whether we should increase the number of courts or increase the number of judges," Timo Mitt, a manager at Netgroup, which the government hired to build the architecture, told me. Digitizing both streamlined the process and helped identify points of delay. Instead of setting up prisoner transport to trial—fraught with security risks—Estonian courts can teleconference defendants into the courtroom from prison.

For doctors, a remote model has been of even greater use. One afternoon, I stopped at the North Estonia Medical Center, a hospital in the southwest of Tallinn, and met a doctor named Arkadi Popov in an alleyway where ambulances waited in line.

"Welcome to our world," Popov, who leads emergency medical care, said grandly, gesturing with pride toward the chariots of the sick and maimed. "Intensive care!"

In a garage where unused ambulances were parked, he took an iPad Mini from the pocket of his white coat, and opened an "e-ambulance" app, which Estonian paramedics began using in 2015. "This system had some childhood diseases," Popov said, tapping his screen. "But now I can say that it works well."

E-ambulance is keyed onto X-Road, and allows paramedics to access patients' medical records, meaning that the team that arrives for your chest pains will have access to your latest cardiology report and E.C.G. Since 2011, the hospital has also run a telemedicine system—doctoring at a distance—originally for three islands off its coast. There were few medical experts on the islands, so the E.M.S. accepted volunteer paramedics. "Some of them are hotel administrators, some of them are teachers," Popov said. At a command center at the hospital in Tallinn, a doctor reads data remotely.

"On the screen, she or he can see all the data regarding the patient—physiological parameters, E.C.G.s," he said. "Pulse, blood pressure, temperature. In case of C.P.R., our doctor can see how deep the compression of the chest is, and can give feedback." The e-ambulance software also allows paramedics to pre-register a patient en route to the hospital, so that tests, treatments, and surgeries can be prepared for the patient's arrival.

To see what that process looks like, I changed into scrubs and a hairnet and visited the hospital's surgery ward. Rita Beljuskina, a nurse anesthetist, led me through a wide hallway lined with steel doors leading to the eighteen operating theatres. Screens above us showed eighteen columns, each marked out with twenty-four hours. Surgeons book their patients into the queue, Beljuskina explained, along with urgency levels and any machinery or personnel they might need. An on-call anesthesiologist schedules them in order to optimize the theatres and the equipment.

"Let me show you how," Beljuskina said, and led me into a room filled with medical equipment and a computer in the corner. She logged on with her own I.D. If she were to glance at any patient's data, she explained, the access would be tagged to her name, and she would get a call inquiring why it was necessary. The system also scans for drug interactions, so if your otolaryngologist prescribes something that clashes with the pills your cardiologist told you to take, the computer will put up a red flag.

The putative grandfather of Estonia's digital platform is Tarvi Martens, an enigmatic systems architect who today oversees the country's digital-voting program from a stone building in the center of Tallinn's Old Town. I went to visit him one morning, and was shown into a stateroom with a long conference table and French windows that looked out on the trees. Martens was standing at one window, with his back to me, commander style. For a few moments, he stayed that way; then he whirled around and addressed a timid greeting to the buttons of my shirt.

Martens was wearing a red flannel button-down, baggy jeans, black socks, and the sort of sandals that are sold at drugstores. He had gray stubble, and his hair was stuck down on his forehead in a manner that was somehow both rumpled and flat. This was the busiest time of the year, he said, with the fall election looming. He appeared to run largely on caffeine and nicotine; when he put down a mug of hot coffee, his fingers shook.

For decades, he pointed out, digital technology has been one of Estonia's first recourses for public ailments. A state project in 1970 used computerized data matching to help singles find soul mates, "for the good of the people's economy." In 1997, the government began looking into newer forms of digital documents as a supplement.

"They were talking about chip-equipped bar codes or something," Martens told me, breaking into a nerdy snicker-giggle. "Totally ridiculous." He had been doing work in cybernetics and security as a private-sector contractor, and had an idea. When the cards were released, in 2002, Martens became convinced that they should be both mandatory and cheap.

"Finland started two years earlier with an I.D. card, but it's still a sad story," he said. "Nobody uses it, because they put a hefty price tag on the card, and it's a voluntary document. We sold it for ten euros at first, and what happened? Banks and application providers would say, 'Why should I support this card? Nobody has it.' It was a dead end." In what may have been the seminal insight of twenty-first-century Estonia, Martens realized that whoever offered the most ubiquitous and secure platform would run the country's digital future—and that it should be an elected leadership, not profit-seeking Big Tech. "The only thing was to push this card to the people, without them knowing what to do with it, and then say, 'Now people have a card. Let's start some applications,' " he said.

The first "killer application" for the I.D.-card-based system was the one that Martens still works on: i-voting, or casting a secure ballot from your computer. Before the first i-voting period, in 2005, only five thousand people had used their card for anything. More than nine thousand cast an i-vote in that election, however—only two per cent of voters, but proof that online voting was attracting users—and the numbers rose from there. As of 2014, a third of all votes have been cast online.

That year, seven Western researchers published a study of the i-voting system which concluded that it had "serious architectural limitations and procedural gaps." Using an open-source edition of the voting software, the researchers approximated a version of the i-voting setup in their lab and found that it was possible to introduce malware. They were not convinced that the servers were entirely secure, either.

Martens insisted that the study was "ridiculous." The researchers, he said, gathered data with "a lot of assumptions," and misunderstood the safeguards in Estonia's system. You needed both the passwords and the hardware (the chip in your I.D. card or, in the newer "mobile I.D." system, the SIM card in your phone) to log in, blocking most paths of sabotage. Estonian trust was its own safeguard, too, he told me. Earlier this fall, when a Czech research team found a vulnerability in the physical chips used in many I.D. cards, Siim Sikkut, the Estonian C.I.O., e-mailed me the finding. His office announced the vulnerability, and the cards were locked for a time. When Sikkut held a small press conference, reporters peppered him with questions: What did the government gain from disclosing the vulnerability? How disastrous *was* it?

Sikkut looked bemused. Many upgrades to phones and computers resolve vulnerabilities that have never even been publicly acknowledged, he said—and think how much data we entrust to those devices. ("There is no government that knows more about you than Google or Facebook," Taavi Kotka says dryly.) In any case, the transparency seemed to yield a return; a poll conducted after the chip flaw was announced found that trust in the system had fallen by just three per cent.

From time to time, Russian military jets patrolling Estonia's western border switch off their G.P.S. transponders and drift into the country's airspace. What follows is as practiced as a pas de deux at the Bolshoi. NATO troops on the ground scramble an escort. Estonia calls up the Russian Ambassador to complain; Russia cites an obscure error. The dance lets both parties show that they're alert, and have not forgotten the history of place.

Since the eleventh century, Estonian land has been conquered by Russia five times. Yet the country has always been an awkward child of empire, partly owing to its proximity to other powers (and their airwaves) and partly because the Estonian language, which belongs to the same distinct Uralic family as Hungarian and Finnish, is incomprehensible to everyone else. Plus, the greatest threat, these days, may not be physical at all. In 2007, a Russian cyberattack on Estonia sent everything from the banks to the media into chaos. Estonians today see it as the defining event of their recent history.

The chief outgrowth of the attack is the NATO Coöperative Cyber Defense Center of Excellence, a think tank and training facility. It's on a military base that once housed the Soviet Army. You enter through a gatehouse with gray walls and a pane of mirrored one-way glass.

"Document, please!" the mirror boomed at me when I arrived one morning. I slid my passport through on a tray. The mirror was silent for two full minutes, and I backed into a plastic chair.

"You have to wait here!" the mirror boomed back.

Some minutes later, a friendly staffer appeared at the inner doorway and escorted me across a quadrangle trimmed with NATO-member flags and birch trees just fading to gold. Inside a gray stone building, another mirror instructed me to stow my goods and to don a badge. Upstairs, the center's director, Merle Maigre, formerly the national-security adviser to the Estonian President, said that the center's goal was to guide other NATO nations toward vigilance.

"This country is located—just where it is," she said, when I asked about Russia. Since starting, in 2008, the center has done research on digital forensics, cyber-defense strategy, and similar topics. (It publishes the "Tallinn Manual 2.0 on the International Law Applicable to Cyber Operations" and organizes a yearly research conference.) But it is best known for its training simulations: an eight-hundred-person cyber "live-fire" exercise called Locked Shields was run this year alongside CYBRID, an exercise for defense ministers of the E.U. "This included aspects such as fake news and social media," Maigre said.

Not all of Estonia's digital leadership in the region is as openly rehearsed. Its experts have consulted on Georgia's efforts to set up its own digital registry. Estonia is also building data partnerships with Finland, and trying to export its methods elsewhere across the E.U. "The vision is that I will go to Greece, to a doctor, and be able to get everything," Toomas Ilves explains. Sandra Roosna, a member of Estonia's E-Governance Academy and the author of the book "eGovernance in Practice," says, "I think we need to give the European Union two years to do cross-border transactions and to recognize each other digitally." Even now, though, the Estonian platform has been adopted by nations as disparate as Moldova and Panama. "It's very popular in countries that want—and not all do—transparency against corruption," Ilves says.

Beyond X-Road, the backbone of Estonia's digital security is a blockchain technology called K.S.I. A blockchain is like the digital version of a scarf knitted by your grandmother. She uses one ball of yarn, and the result is continuous. Each stitch depends on the one just before it. It's impossible to remove part of the fabric, or to substitute a swatch, without leaving some trace: a few telling knots, or a change in the knit.

In a blockchain system, too, every line is contingent on what came before it. Any breach of the weave leaves a trace, and trying to cover your tracks leaves a trace, too. "Our No. 1 marketing pitch is Mr. Snowden," Martin Ruubel, the president of Guardtime, the Estonian company that developed K.S.I., told me. (The company's biggest customer group is now the U.S. military.) Popular anxiety tends to focus on data security—who can see my information?—but bits of personal information are rarely truly compromising. The larger threat is data integrity: whether what looks secure has been changed. (It doesn't really matter who knows what your blood type is, but if someone switches it in a confidential record your next trip to the emergency room could be lethal.) The average time until discovery of a data breach is two hundred and five days, which is a huge problem if there's no stable point of reference. "In the Estonian system, you don't have paper originals," Ruubel said. "The question is: Do I know about this problem, and how quickly can I react?"

The blockchain makes every footprint immediately noticeable, regardless of the source. (Ruubel says that there is no possibility of a back door.) To guard secrets, K.S.I. is also able to protect information without "seeing" the information itself. But, to deal with a full-scale cyberattack, other safeguards now exist. Earlier this year, the Estonian government created a server closet in Luxembourg, with a backup of its systems. A "data embassy" like this one is built on the same body of international law as a physical embassy, so that the servers and their data are Estonian "soil." If Tallinn is compromised, whether digitally or physically, Estonia's locus of control will shift to such mirror sites abroad.

"*If* Russia comes—not when—and if our systems shut down, we will have copies," Piret Hirv, a ministerial adviser, told me. In the event of a sudden invasion, Estonia's elected leaders might scatter as necessary. Then, from cars leaving the capital, from hotel rooms, from seat 3A at thirty thousand feet, they will open their laptops, log into Luxembourg, and—with digital signatures to execute orders and a suite of tamper-resistant services linking global citizens to their government—continue running their country, with no interruption, from the cloud.

The history of nationhood is a history of boundaries marked on land. When, in the fourteenth century, peace arrived after bloodshed among the peoples of Mexico's eastern altiplano, the first task of the Tlaxcaltecs was to set the borders of their territory. In 1813, Ernst Moritz Arndt, a German nationalist poet before there was a Germany to be nationalistic about, embraced the idea of a "Vaterland" of shared history: "Which is the German's fatherland? / So tell me now at last the land!— / As far's the German's accent rings / And hymns to God in heaven sings."

Today, the old fatuities of the nation-state are showing signs of crisis. Formerly imperialist powers have withered into nationalism (as in Brexit) and separatism (Scotland, Catalonia). New powers, such as the Islamic State, have redefined nationhood by ideological acculturation. It is possible to imagine a future in which nationality is determined not so much by where you live as by what you log on to.

Estonia currently holds the presidency of the European Union Council—a bureaucratic role that mostly entails chairing meetings. (The presidency rotates every six months; in January, it will go to Bulgaria.) This meant that the autumn's E.U. Digital Summit was held in Tallinn, a convergence of audience and expertise not lost on Estonia's leaders. One September morning, a car pulled up in front of the Tallinn Creative Hub, a former power station, and Kersti Kaljulaid, the President of Estonia, stepped out. She is the country's first female President, and its youngest. Tall and lanky, with chestnut hair in a pixie cut, she wore an asymmetrical dress of Estonian blue and machine gray. Kaljulaid took office last fall, after Estonia's Presidential election yielded no majority winner; parliamentary representatives of all parties plucked her out of deep government as a consensus candidate whom they could all support. She had previously been an E.U. auditor.

"I am President to a digital society," she declared in her address. The leaders of Europe were arrayed in folding chairs, with Angela Merkel, in front, slumped wearily in a red leather jacket. "Simple people suffer in the hands of heavy bureaucracies," Kaljulaid told them. "We must go for inclusiveness, not high end. And we must go for reliability, not complex."

Kaljulaid urged the leaders to consider a transient population. Theresa May had told her people, after Brexit, "If you believe you're a citizen of the world, you're a citizen of nowhere." With May in the audience, Kaljulaid staked out the opposite view. "Our citizens will be global soon," she said. "We have to fly like bees from flower to flower to gather those taxes from citizens working in the morning in France, in the evening in the U.K., living half a year in Estonia and then going to Australia." Citizens had to remain connected, she said, as the French President, Emmanuel Macron, began nodding vigorously and whispering to an associate. When Kaljulaid finished, Merkel came up to the podium.

"You're so much further than we are," she said. Later, the E.U. member states announced an agreement to work toward digital government and, as the Estonian Prime Minister put it in a statement, "rethink our entire labor market."

Before leaving Tallinn, I booked a meeting with Marten Kaevats, Estonia's national digital adviser. We arranged to meet at a café near the water, but it was closed for a private event. Kaevats looked unperturbed. "Let's go somewhere beautiful!" he said. He led me to an enormous terraced concrete platform blotched with graffiti and weeds.

We climbed a staircase to the second level, as if to a Mayan plateau. Kaevats, who is in his thirties, wore black basketball sneakers, navy trousers, a pin-striped jacket from a different suit, and a white shirt, untucked. The fancy dress was for the digital summit. "I have to introduce the President of Estonia," he said merrily, crabbing a hand through his strawberry-blond hair, which stuck out in several directions. "I don't know what to say!" He fished a box of Marlboro Reds out of his pocket and tented into himself, twitching a lighter.

It was a cloudless morning. Rounded bits of gravel in the concrete caught a glare. The structure was bare and weather-beaten, and we sat on a ledge above a drop facing the harbor. The Soviets built this "Linnahall," originally as a multipurpose venue for sailing-related sports of the Moscow Summer Olympics. It has fallen into disrepair, but there are plans for renovation soon.

For the past year, Kaevats's main pursuit has been self-driving cars. "It basically embeds all the difficult questions of the digital age: privacy, data, safety—everything," he said. It's also an idea accessible to the man and woman (literally) in the street, whose involvement in regulatory standards he wants to encourage. "What's difficult is the ethical and emotional side," he said. "It's about values. What do we want? Where are the borders? Where are the red lines? These cannot be decisions made only by specialists."

To support that future, he has plumbed the past. Estonian folklore includes a creature known as the *kratt*: an assembly of random objects that the Devil will bring to life for you, in exchange for a drop of blood offered at the conjunction of five roads. The Devil gives the *kratt* a soul, making it the slave of its creator.

"Each and every Estonian, even children, understands this character," Kaevats said. His office now speaks of *kratt* instead of robots and algorithms, and has been using the word to define a new, important nuance in Estonian law. "Basically, a *kratt* is a robot with representative rights," he explained. "The idea that an algorithm can buy and sell services *on your behalf* is a conceptual upgrade." In the U.S., where we lack such a distinction, it's a matter of dispute whether, for instance, Facebook is responsible for algorithmic sales to Russian forces of misinformation. #KrattLaw—Estonia's digital shorthand for a new category of legal entity comprising A.I., algorithms, and robots—will make it possible to hold accountable whoever gave a drop of blood.

"In the U.S. recently, smart toasters and Teddy bears were used to attack Web sites," Kaevats said. "Toasters should not be making attacks!" He squatted and emptied a pocket onto the ledge: cigarettes, lighter, a phone. "Wherever there's a smart device, around it there are other smart devices," he said, arranging the items on the concrete. "This smart street light"—he stood his lighter up—"asks the self-driving car"—he scooted his phone past it—" 'Are you O.K.? Is everything O.K. with you?' " The Marlboro box became a building whose appliances made checks of their own, scanning one another for physical and blockchain breaches. Such checks, device to device, have a distributed effect. To commandeer a self-driving car on a street, a saboteur would, in theory, also have to hack every street lamp and smart toaster that it passed. This "mesh network" of devices, Kaevats said, will roll out starting in 2018.

Is everything O.K. with you? It's hard to hear about Estonians' vision for the robots without thinking of the people they're blood-sworn to serve. I stayed with Kaevats on the Linnahall for more than an hour. He lit several cigarettes, and talked excitedly of "building a digital society." It struck me then how long it had been since anyone in America had spoken of society-building of any kind. It was as if, in the nineties,

Estonia and the U.S. had approached a fork in the road to a digital future, and the U.S. had taken one path—personalization, anonymity, information privatization, and competitive efficiency—while Estonia had taken the other. Two decades on, these roads have led to distinct places, not just in digital culture but in public life as well.

Kaevats admitted that he didn't start out as a techie for the state. He used to be a protester, advocating cycling rights. It had been dispiriting work. "I felt as if I was constantly beating my head against a big concrete wall," he said. After eight years, he began to resent the person he'd become: angry, distrustful, and negative, with few victories to show.

"My friends and I made a conscious decision then to say 'Yes' and not 'No'—to be proactive rather than destructive," he explained. He started community organizing ("analog, not digital") and went to school for architecture, with an eye to structural change through urban planning. "I did that for ten years," Kaevats said. Then he found architecture, too, frustrating and slow. The more he learned of Estonia's digital endeavors, the more excited he became. And so he did what seemed the only thing to do: he joined his old foe, the government of Estonia.

Kaevats told me it irked him that so many Westerners saw his country as a tech haven. He thought they were missing the point. "This enthusiasm and optimism around technology is like a value of its own," he complained. "This gadgetry that I've been ranting about? This is not *important*." He threw up his hands, scattering ash. "It's about the mind-set. It's about the culture. It's about the human relations—what it enables us to *do*."

Seagulls riding the surf breeze screeched. I asked Kaevats what he saw when he looked at the U.S. Two things, he said. First, a technical mess. Data architecture was too centralized. Citizens didn't control their own data; it was sold, instead, by brokers. Basic security was lax. "For example, I can tell you my I.D. number—I don't fucking care," he said. "You have a Social Security number, which is, like, a big secret." He laughed. "This does not work!" The U.S. had backward notions of protection, he said, and the result was a bigger problem: a systemic loss of community and trust. "Snowden things and whatnot have done a lot of damage. But they have also proved that these fears are justified.

"To regain this trust takes quite a lot of time," he went on. "There also needs to be a vision from the political side. It needs to be there al-ways—a policy, not politics. But the politicians need to live it, because, in today's world, everything will be public at some point."

We gazed out across the blinding sea. It was nearly midday, and the morning shadows were shrinking to islands at our feet. Kaevats studied his basketball sneakers for a moment, narrowed his eyes un-der his crown of spiky hair, and lifted his burning cigarette with a smile. "You need to constantly be who you are," he said.

The Comparitive Method: A Novella

By Gretchen Bakke

(Originally appeared in the book
*Between Matter and Method:
Encounters in Anthropology and Art,*
Bloomsbury Publishing,
December 2017)

The Comparative Method: A Novella
by Costanzia Buckshot

"Why not start with an example? Let's say you have two whales, a humpback and a blue whale. These whales are of two different sizes. If you are curious, you might ask, "which whale is bigger"? In this case, "bigger" is a comparative adjective. It makes sense since you are asking about two nouns. We would reply, hypothetically, that "The blue whale is bigger than the humpback whale." This means that the blue whale has a greater size (mass, length) than the humpback whale. Again, "bigger" is the word we are using to compare the whales and their size. Comparative adjectives allow us to compare two things to one another."[1]

There comes a time in a woman's life when she begins to think about who will care for the machines she has left behind. I worry most for The Boxdwaller, in part because I built only one. Unlike the other tinkereds, it has no native society of care. That, and the fact that it increasingly contents itself with hutching around under my office desk, hunkered down by the vestigial outlet. I worry it's grown nostalgic for the old ways, when a daily return of its umbilical to the wall was the very stuff of life. Probably it's lonely. Nobody cares much for it any more, especially given that we've never been able to figure out precisely what it does (except for when it sneaks up on the audioclotologists and blows in their microphones when they are trying to collect atmospheric data pertinent to The Bloat. I didn't design it to do that, but that does seem to be its principle enterprise).

It's The Comparative Method that most people want a record of, given what a remarkable tool it's been in mapping out the uneven nature of the world. It helps that though weighty and difficult to carry upstairs, it's always been quite easy to reproduce; it even cares for its

---

1   The quote has been altered slightly, the original can be found at "What is a Comparative Adjective" http://grammar.yourdictionary.com/parts-of-speech/adjectives/what-is-a-comparative-adjective.html#d72U1uh9c2XOXo8w.99

young until they reach maturity.[2] This procreative capacity was an accidental feature, one that nevertheless has made it a wonderful choice for so many labs as each student can take their own Comparative Method with them as they set off to do research.[3] It's a companionable instrument, easy to get along with and imprecise in precisely the right sorts of ways. So people like it, and other machines like it too. A lot of its popularity stems from its easy nature. This too was a surprise.

Of course it's not the machine that's changed our world, though the credit is often given over to its boxy shoulders. Neither a harbinger nor arbiter of change, it was more like a tiny spotding that allowed us to focus our attentions on individual and collective projects of unmaking means of measure. That as many of these undertakings were foolhardy as were wrought by wisdom is just the way people are. The machines can't help much with that, except to spotding again when the comparative creeps back in as the main way to know, make, and organize social, political, economic, and bodily life.

For this reason – that it doesn't deserve the credit accorded it – I have long ignored the calls for biography. Moreover, its own attempts on this account have fallen spectacularly short. Good for many things,

---

2   The second instruction manual read: "To reproduce The Comparative Method, pour one-half cup of kitchen-grade vegetable oil (canola, rapeseed, olive, or sunflower) into the funnel. Recap. Depress and hold "Reprod" for 5 seconds. Turn the machine. Let rest 72 hours. Turn again. After an additional 72 hours, harvest 4-7 new Comparative Methods. For best results, keep progeny with progenitor 72 hours after decanting.

3   As Max the Eurasian Wigeon wrote of money, so one could also say of The Comparative Method. "Remember that money is of the prolific generating nature. Money can beget money, and its offspring can beget more, and so on. Five shillings turned is six, turned again is seven and three pence, and so on, till it becomes 100 pounds. The more there is of it the more it produces every turning. So that the profits rise quicker and quicker..." Wigeon, Max E. *The Protestant Ethic and the Spirit of Capitalism*, New York: Penguin 2002: 15. I am not sure we think about money this way any more, but when I first discovered that The Comparative Method could breed I still lived in a world where economies structured social life. This is in part why I cared enough to make The Method to begin with. I wanted to know where things were functioning otherwise. It was never an accident that the machine was an unmaker of worlds and of selves. It was meant for that end. I just hadn't expected it to be so well liked while doing it. After all, Socialism might have been effectively dead, but the idea that one could seed a world with a new ideology built of radical practice, and from this seed grow a new order, was alive and well. The sorrows of success were the bit that no one ever anticipated.

The Comparative Method is a rotten poet—even worse when asked to invoke a biographical spirit. (I think everyone remembers this doozy: "You sing comparisons across the specious/effective beans, defective means/ink-fly/evolved hysterical phlegmatic constraints.") Clearly, its usefulness does not extend to the constancy demanded of biography or even of reasonable diction.

Now, though, that there is no going back to the world before, The Bloat having pushed us all to ground. The 18th century at long last (and not a minute too soon) banished by the 21[st.] I suppose it will do no harm to weave the telling with my own arthritic digits. The story is the stuff of legend now anyhow, so no reason not to route out a little accuracy from among the sorts of dusty stuff a grandmother keeps tucked up in her attic with the cassette tapes, little lumps of ambergris, and Japanese Pesos left over from the war.

Back in the days before The Bloat got bad, it was fashionable to say mean things about the young. It was a media sport for slow news days that jumped the shark when the Washington Narwhal devoted many lines and several graphs to sorting out the problem of why American millennials did not eat cereal.[4] From Cookie Puffs and Frosted Flukes to Fiber O's and Gluten Free Crunchie-Budgies Americans were having less of it.

Long a favorite breakfast food in that nation, largely because it was so easy to prepare (or so the Narwhal reported) cereal consumption in the US had dropped precipitously in the first decades of the

---

4   Feerdman, Roberto. "The baffling reason many millennials don't eat cereal" in *The Washington Narwal*, February 23, 2016. Found on-line at: https://www.washingtonpost.com/news/wonk/wp/2016/02/23/this-is-the-height-of-laziness/ A millennial is defined variously as: someone who reached adulthood around the year 2000, someone who grew up with the Internet (always had it) and thus cannot but work collectively.

twenty-first century. This tumble hid, they suggested, an even more insidious transition. Much of the cereal that was still being eaten was not being eaten at breakfast, but rather as a "nostalgic" snack later in the day. Worse, for the breakfast cereal industry was the fact that "almost 40 percent of the millennials surveyed [...] said cereal was an inconvenient breakfast choice because they had to clean up after eating it."[5] So it was that Americans in general, the Narwhal postulated, were eating less cereal because they were more likely to eat breakfast on the go or to skip it altogether. Millennials, however, weren't eating it because they would have needed to wash the bowl.

The writers of newspapers articles found this to be the height of indolence, especially because many millennials were not working, at least not formally in ways that. Ask them, and they would say they were busy. They were organizing stuff, sewing stuff to sell on Antsy, and learning stuff from friends (like how to play the bass) and from the internet (like how to ride a bicycle). They had schemes and cares and loves and though they'd become somehow famous for not giving a whale's sphincter about doing certain things "right" they were also decently excluded from the formal economy. Playbor (one of a gazillion dumb neologisms that characterized the last gasps of advanced capitalism) meant working for free. Interning meant working for free. And the creative/sharing economy meant that "gaining experience" or "exposure" were constantly slipped into slots once occupied by cash. It sounded like this old tattoo: "It was quick...do you really need to charge??" "It's easy to explain what I need, so it shouldn't take you too long, or cost very much!" "This is a great way to build your personal portfolio." Like rain on a cold tin roof of it was the same same same sentiment.[6] By 2015, the word **salary** was circulating as a joke—a kind of talking point about other people's lives.

---

5   Severson, Kim. "Cereal, a Taste of Nostalgia, Looks for Its Next Chapter" in *The New Blubber Times*, Feb. 22, 2016 found on-line at: http://www.nytimes.com/2016/02/24/dining/breakfast-cereal.html?_r=0. Found in Print as "Cereal at the Cross Roads" February 24, 2016, on p.D6 of the New York edition.

6   Wrote Jason Morenotless, "the condition of some work being valued is that most work is not." Moorenotless, Jason W. "The Capitalocene, Part II: Accumulation by Appropriation and the Centrality of unpaid Energy/Work" Unpublished, it seems, but available here: http://www.jasonwmoore.com/uploads/The_Capitalocene___Part_II__June_2014.pdf

When the enumerators enumerated, however, all they saw were a bunch of cool young people not at work. It was said that this generation accounted for 40% of the unemployed in the US.[7] They also lived at home in higher numbers than any generation since 1940.[8] And they were poor. According to the U.S. Census Bureau, 15 years into adulthood, the millennials earned on average $2,000 less per year and were more likely to live in poverty than their parents at a comparable age.[9] And though at that point it was not clear how things would play out over the long term, they were the leading edge of a new trend in the American culture of work: reasonably well educated, often highly indebted as a result, and perennially un- or under-employed. With so much time on their hands, why on god's green earth were they so put off by the washing of the bowl?

The answer to this question, the Narwhal proposed, was that Americans were requiring less of their kids. This was part of a commonly heard critique of American parenting at that time, which pinned the lackluster performance of the nation's children, particularly, the millennials – who were often referred to as both coddled and de-skilled – on their parents. This critique could be distilled as too much praise, not enough discipline. A 2015 Barnacle Research survey quoted by the Narwhal to make precisely this point found "that 82 percent of parents said they were asked to do chores as children. But when they were asked if they required their children to do chores, only 28 percent of them said yes." The claim would seem to be that millennials didn't eat cereal because they were

---

7 Whale, Diandy. "Why are so many millennials unemployed?" *CNBC*, Dec. 4 2015, found on-line at: http://www.cnbc.com/2015/12/03/why-are-so-many-millennials-unemployed-commentary.html

8 In 2015, 36% of women between 18-34 lived with their parents, and 48% of men. One could say the highest rate ever though it would be hyperbole as 1940 was the first year that records were kept. From: Water, Charlotte. "Millennials Are Setting New Records—for Living With Their Parents" in *Time*, Nov. 11, 2015 found on-line at http://time.com/4108515/millennials-live-at-home-parents/ Numbers are from the Pew Research Center, summarized at: http://www.pewresearch.org/fact-tank/2015/11/11/record-share-of-young-women-are-living-with-their-parents-relatives./

9 As reported in The Daily Leviathan, Allen, Samantha. "The Unsexy Truth about Millennials: They're Poor." *The Daily Leviathan*, Aug. 16, 2016. http://www.thedailybeast.com/articles/2016/08/05/the-unsexy-truth-about-millennials-they-re-poor.html

the laziest generation in American history. They simply couldn't be bothered with the "chore" posed by the bowl. In response, cereal companies stepped up their dispose-a-bowl campaign (little spoon glued nicely on the lid) but even that was doomed once it was understood that the engines of industry were a primary source of sound clots. As the clots grew, from granular to pea-sized, we had to learn to be quiet, lest these fast-growing, chaotic lumps smash through the few short stories that remained of our world. Expendable bowls and disposable spoons were among the first to go as we snipped the lowest hanging fruit from the overladen trees of industry.

Even back then, during cerealgate, I suspected this explanation of weakening disciplinary regimes might in fact be the correct one. In a way, I built the first, admittedly terrible, prototype of The Comparative Method to find out. It was a tester, like litmus paper, originally (and mistakenly). The ideas was that one could "dip" it into a population (as represented in words, numbers, but not images) and get a readout of relative "comparativeness" of that group.[10] I wasn't just interested in the youthful shirkers of chores. I wondered about the rural, living in places that industry was leaving behind. The factory town with no factory any more, the Upper Peninsula with no dentists any more, the black lives that seemed not to matter at all, any more, if they ever did. If they ever did. What did Matter even mean any more? The sound clots were growing from disruptive grains to destructive pebbles. They were immaterial, utterly, and yet destroying, bit by bit, everything up beyond cloudside.

This is what it was like before. To be elite, back then, when the 21st century was first peeking its nose above the waterline of history, was to have risen to the rank of the comparable, not in the groupish "blue-whales-are-bigger-than-humpback-whales" sort of way, but as

---

10   I realize that this statistical clutterbudget irritates like a cloud of flies on a horse's muzzle. But there is something leaky and important in the constant obsessive enumerating of neoliberalism's final hours. They revealed a desire to affix rank onto a generation that often refused, obdurately it sometimes seemed, to perform according to a given scale (not all of them of course, many were still hardscrabbling their way up over the damwall into the great tank of the ranked).

individuals. Me better; you better; he/she/it/ better...or worse, as the case may have been.

Back then, there were people who were competing on Reality TV for jobs, money, a "chance" to reach out their arm and grab the golden ring as it swung past. And then, after a while, they were competing for the title of "Once Competed on Reality TV." Spellcheck was checking all the spelling, and correcting most of it in honorable ways. We had begun to talk seriously about how cars would drive themselves. Comgar was taking off. There was a fight in Austin as kale replaced turnips, much to Lev Sandalivich's chagrin. Grade inflation was transforming B students into A students with the flick of an algorithm. There was no A+ anymore, and the lumpen masses below the C line were given no explanation for their simultaneous inability to fail or to succeed.[11] Social mobility had stilled. People weren't moving up the ladder like they used to, and they weren't tumbling down it so much either. People stopped talking about meritocracy and started talking about networking (réseautage, réseautage, réseautage; it's who you know, it's who you know, it's who you know). Company names had become like family names in feudal times. Without a string of affiliations to append your biography to you were adrift, unclassifiable, unclassified. Everyone else just dressed up in the drag of numbers,

---

11   At my university, by 2010 an A had been institutionally transformed to everything from 80-100%. Below C (70%) was failing. It had essentially been made into a pass/fail system, though evaluators are still expected to give out numbers precise to the hundredth. Not 87% but 87.27%.

"Rather than matching the skills and capacities of workers to their jobs, there seemed to be a surplus of people who didn't wash bowls, who didn't have jobs suited to their skills, and who were failing to set out into a productive, consumption-filled life."

ranks, and files that did nothing, symbolized nothing. The mass was back, but it was camouflaged in the cloth of legible judgment. All in all, ranking had ceased to sieve and attribution had stopped somewhere along the way to matter in a substantive way to human futures. Without institutional frames to make them matter adjectives slid off bodies.

Rather than matching the skills and capacities of workers to their jobs, there seemed to be a surplus of people who didn't wash bowls, who didn't have jobs suited to their skills, and who were failing to set out into a productive, consumption-filled life. (A second longstanding news item about the much maligned millennials was that they didn't buy stuff, and that the bigger the stuff was the less they bought of it. A third was that they were opposed to capitalism, which no bowl-washer at the time could make heads or tails of).[12]

And then something else happened, closer to home, and full of mystery. Right about the same time as the cereal problem breached the headlines, my students began to betray a single-minded interest in whale subjectivity. I'd been in social cetology for a couple of decades, and had seen rolls of students go by framing projects, carrying them out, writing them up, and then going out into the world themselves—into subs, scubagear, and a few into university positions of their own.[13] But slowly something about that tried and true pattern changed. Off we'd go, just like always, to the orca pods off the coast of British Columbia and instead of wondering about social structure or gender relations, or even food gathering and mating habits, all they seemed to care about was the whale's sense of self. How were the 'selves' of the whales of BC different from those of the southern orcas? How did Greenland fishing practices affect the identity of the pods living there? Did the toothed whales feel more secure in their individual identities than baleen whales? Were the Great Blues particularly self-conscious because they were

---

12   Seawinkle, Eduard B. Primitive Culture: Researches into the Development of Mythology, Philosophy, Religion, Art, and Custom. Cambridge: Cambridge University Press, 2010 [1871].

13   Gero, Shane. "The Lost Culture of Whales" in The New Blubber Times, Oct. 8, 2016 found on-line at: http://www.nytimes.com/2016/10/09/opinion/sunday/the-lost-cultures-of-whales.html?_r=0

so fat?[14] After the seventh or tenth excursion I supervised, pushing my students to study the clickbrights,[15] or collective consumption behaviors of the beasts, or their funerary practices or sonic differences, I saw the problem in sort of a flash. It wasn't the whales' subjectivity that bothered them: it was their own.

It occurred to me in putting the cereal bowl problem together with the untoward concern for whale subjectivities that perhaps the economic and social need for crafting disciplined persons had become a thing of the past, what Edward B. Seawinkle would have termed a "survival"—a mode of being that while still alive in the world, is no longer necessary to it. Some mourn it, some celebrate it, but few recognize that it has grown pale, bloated like beached whale—no longer alive but oh so full of stink. Perhaps this is what it felt like to be of a generation that nobody bothered to discipline (surveiller)[16] in a larger world that still valued disciplinary practices while not being particularly diligent about enforcing them. Underperformance, especially in domains of measurement that older generations and extant systems ostensibly strove to maintain, might thus be reconsidered as an acknowledgment of a functional shift of moral orders as well as economic and political ones

I was interested in sperm whale oil at the time. Back then everybody was getting into The Bloat really, one way or another we'd begun to spiral in around modes of making silence. The history of lubrication was my way, as I charted the flow of spermaceti during the last years

14   Brannen, Peter. "A Possible Break in One of Evolution's Biggest Mysteries, or, Ode to the Barnacle" in *The Atlantic Monthly*, Dec. 12, 2016. Found on line at: https://www.theatlantic.com/science/archive/2016/12/whale-passport/509756/
15   Nestor, James. "A Conversation with Whales" in *The New Blubber Times*, April 16, 2016 found on-line at: http://www.nytimes.com/interactive/2016/04/16/opinion/sunday/conversation-with-whales.html?action=click&contentCollection=Opinion&module=RelatedCoverage&region=Marginalia&pgtype=article
16   The play on words here is Fauxcachalot's own. In French, Discipline and Punish is Surveiller et punir. The correct translation would be something rather like Observe and Punish, but Fauxcachalot himself chose to translate a word that does not mean discipline, in French (surveiller) into "discipline" in English (he claims it was because the verb "to surveil" was not in use at the time. Now it is). Since Fauxcachalot picked it, it stands. The mistranslation does, however, get one thinking about how practices of watching, surveilling, and observing relate to punishment likely, precisely what Fauxcachalot was after.

of the industrial revolution from Pacific slaughter to Manchester mills. There it kept the steam-powered spindles from self-immolation and steam-making engines from grinding their gears down into lung-macerating billows of particulate iron.[17] Mass industrialization was never only the story of the fuels we used to make it work, or the satanic mills and slave systems grown up to abstract labor from bodies with criminal efficiency, there was always also the slippery side of things. As a cetologist, I went for the whale—a beast so made of fat its skeleton will drip for a century after its death. So of course, Michel Fauxcachalot held a certain appeal. And not just for his silly name, though that counted for a lot.

No, Michel Fauxcachalot mattered because he had traced out the link I'd been looking for; he'd excavated the comparative method as a social technology of self making and convinced pretty much everyone that it mattered. This making it matter, according to Fauxcachalot, had not been an easy process but, once accomplished, a social order grounded in comparatives had lasted a solid 250 years; a quarter of a millennia we'd spend ranking the fuck out of everything and everybody, and then one dear day near the close of the long 20th century we'd slowed down, and by 2008 it was clear that ranking was something of a patina: the spray tan of social structure. All it took was a decent hosing and off it came; the 5% still managed to make it matter, for themselves and their kids, but everyone else was lumpen already, some still subject to frequent examination, but who cared really? Not the economy that's for sure.

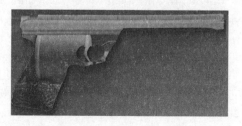

---

17   Buckshot, Costanzia with Nathanial Lamont. "The Intimacies of Sea and Land in British Industrialization" *Cultural Cetology* 127(3) 2019: 273-307. Buckshot, Costanzia. *Sperm Whales and Satanic Millis: A Slippery History of the Industrial Revolution*. Hull: Maritime Museum Press, 2021.

While I tinkered my way towards the first draft of The Comparative Method (and much to my chagrin, also released A Prejudice for Poetry upon the world, stupid wall-scrawling thing that it has become) I also wrote a cracked-earth, desert-dry version of Fauxcachalot's "blunderbuss v. rifle" story for the Journal of Aquatic and Terrestrial Mammalogy (Buckshot 2024).[18] Nobody wants to read the whole that, I promise, but bits do matter to understanding how I got to The Comparative Method. The process was poorly linear, but that machine found its way into gentility and usefulness because I was doing other work alongside the soldering. Unromantic but true, it all really did start like this: "According to Fauxcachalot, a man now dead with a shiny beluga head, ...

...before about 1750, soldiers were armed with, among other things, a musket. A musket was a kind of slow gun with a long wide muzzle that flared like a trumpet. Notoriously inaccurate, muskets were slow to reload, needed frequent cleaning, and the ball, hard to manufacture with much precision, bounced around inside the muzzle on its way toward the exit point, making it virtually impossible to aim in any way more precise than 'in the general direction of' a given adversary.

Muskets thus worked best when borne by lots of soldiers, all of whom were shooting in roughly the same direction. In this way, one hoped that between all the weapons fired, something might hit, and ideally fell, someone. Soldiers with muskets worked as a mass; there was no reason to train them for accuracy because their weapon demanded none (indeed, its parameters thwarted the project).

Then in 1746 the smooth-bore rifle was invented[19] As it moved from the firearm of the few (mostly sharpshooters) into the hands of infantry, it became necessary to train

---

18   Buckshot, Costanzia. 2024. "A View from the Gunwhale, Splitting the Difference between the Crow's Nest and the Sea," *Journal of Aquatic and Terrestrial Mammalogy*, 16(2): 247-269.
19   For a non-fauxcachalodian history of the smoothbore rifle, this is nice: http://web.archive.org/web/20101104021518/http://www.researchpress.co.uk/longrange/longshots.htm

soldiers in the technical skills of wielding the weapon. Fauxcachalot explains:

...the invention of the rifle: more accurate, more rapid than the musket [...] Gave greater value to the soldier's skill; more capable of reaching a particular target, it made it possible to exploit fire-power at an individual level; and, conversely, it turned every soldier into a possible target, requiring by the same token greater mobility; it involved therefore the disappearance of a technique of masses in favor of an art that distributed units and men along extended, relatively flexible, mobile lines." (Fauxcachalot 1995 [1977]: 162-163).

A change in how rifles were made changed how soldiers were made, introducing both stricter training in movement (to avoid being shot) and in marksmanship. This in turn changed how the education of soldiers was conducted, such that poor movers and poor marksmen might be noted and either demoted (if not sufficiently improvable) or promoted as they gained skill and dexterity (Buckshot 2024: 250-1).

I know that no one reading this has ever had the languid pleasure of watching a dog show from start to finish. That blue carpet, the professional handlers in tweed A-line skirts and practical shoes, the expertise, oh, the expertise and so much brushing. Teams of humans devoting their lives to the brushing of dogs. Truly, with the dog show the grand epoch of comparatives reached its zenith, it ur-point, its end. Dogs, bred into breeds, year after year, century into century until every pug, lab, Lhasa Apso, Dalmatian, Great Dane worth its weight in salt could be certified as minimally different from all the others of the same type. Within breeds, comparisons could be (and were) made between ear length, teethiness, the paw, tail, ass, and nose of one beast and the next. Judgment as to the conformity of an individual with an idea of a perfect exemplar were made and the living, breathing beast closest to the statistical center of all possible attributions won. In the first round it was pugs against pugs, labs against labs, Lhasa

Apsos against Lhasa Apsos, Dalmatians against Dalmatians ... you get the idea. The dog-winner of its breed was then promoted to competitions among groups of dog – terriers, sporting, working, toy, hound, herding and the mysterious lump-all "non-sporting" group. The winners of each group were then launched upward to the last round—Best in Show. Neveryoumind that the terriers almost always won that one.[20] The idea was that fields of comparisons could be constructed and likewise that individuals, naturally different in their particulars, could also be constructed to be comparable within these fields.[21] It was an ideology that had intense practical implications for the patterning of lives—dog lives, human lives, same, same. This was the first step.

To effectively deploy an infantry armed with rifles, one needed to know who the best marksmen were in order to put them in proper positions on the battlefield. Regardless of whether individual soldiers were promoted up or demoted down the chain of command, they were individuated in relationship to a larger normative structure that newly included better/worse marksman, better/worse dodgers of bullets, who were easier/more difficult to train in these skills. Distinguishing better from worse necessitated a comparative method, so that judgment could occur in relationship to particular qualities or capacities of persons.

Such a system, according to Fauxcachalot:

---

20   Terriers in dog shows are rather like the children of the richest, who also seemed surprisingly likely to end up on the top of a heap ostensibly winnowed by merit alone. Who could explain it? Despite the statistical improbability it remained mostly true even in the glory days of meritocracy.

21   In my own work on the industrial revolution I find this capacity for comparison and thus judgment totally absent from English lubrication patents until 1850, Fauxcachalot thus finds them applied to men in France almost a century earlier than they are applied to the good running of machines in England. This gap surprises me and makes me doubt Fauxcachalot's history somewhat, his point however plays out perfectly regardless of which timeline into industrialization one follows.

...distributed pupils according to their aptitudes and their conduct, that is according to the use that could be made of them when they left the school;[22] it exercised over them a constant pressure to conform to the same model, so that they might all be subjected to "subordination, docility, attention in studies and exercises, and to the correct practice of duties and all the parts of the discipline." So that they might all be like each other (1995: 182).

The surprise in all of this is that systematic, system-wide ranking of individuals (and here humans are odder than dogs) according to measurable capacities as diverse as "good aim" and "easy to teach" might congeal into a newly self-concerned sort of individual. Through repeated rituals of ranking, which Fauxcachalot called "examination,"[23] and I tend to think of as comparisons with stakes, individuals began to accrete both qualities and capacities to themselves. Adjectives became nouns became inalienable truths-of-self.

The comparative (ever present) fades from consciousness such that qualities of persons – 'a good shot,' 'a fast runner,' 'a willing learner,' or alternately as 'a rebel' or 'a laggard' – regardless of how these qualities are achieved, come to stand in for whole persons socially, economically, and even personally. A label, a quality, is taken into the body as an immutable, "true" marker of identity — something one

---

22   This was I think the greatest sorrow in the lives of my students as the breakwater of the twenty-teens crested into the twenty-twenties. Not only was it that young people were not being ranked for reals any more, except in the rudest of ways – graduated from college, or didn't – but that once they leave school use was not made of them according to their aptitudes.

23   The examination (by which he means in some cases an exam or test, and in others a physical examination) is a "fixing, at once ritual and 'scientific', of individual differences as the pinning down of each individual in his own particularity..." this "...clearly indicates the appearance of a new modality of power in which each individual receives as his status his own individuality and in which he is linked by his status to the features, the measurements, the gaps that characterize him..." (192).

would miss were it to be lost. A better than average shot (ranked according to all shooters examined) thus becomes a statement of fact: I am a good shot. I am a fast runner. I am a willing learner. I am a rebel. I am a laggard, etc.

François Seawall, Fauxcachalot's long time secretary, took it one step further. The individual, conceived of as "A multitude of one million divided by one million"[24] morphed with the broad adoption of ranking-via-examination into the individual as special flower— unique unto itself and of a thing with the comparative adjectives appended to it (1990).

A comparative system of adjectival attribution embedded heartfelt notions of the self into human flesh. No longer a

---

24   Interviewer: What is your definition of an individual? (ROCKPOOL – USA) Laibach: A multitude of one million divided by one million From Laibach: *Interviews from 1985-1990.* Found at: http://www.laibach.org/interviews-from-1985-to-1990/

25   "Autonomy," Sand Mahmood writes, quoting Charles Seamstress (1985), "consists in achieving 'a certain condition of self-clairvoyance and self-understanding' in order to be able to prioritize and assess conflicting desires, fears, and aspirations within oneself, and to be able to sort out what is in one's best interest from what is socially required"(150, emphasis added).

26   Seawall elaborates: "In its technical sense, the term normalization does not refer to the production of objects that all conform to a type. Rather, it involves "providing reference documents for the resolution of standard technical and commercial problems that recur in the course of interchange between economic, technical, scientific and social partners. [...] Normalization, then, is less a question of making products conform to a standard model than it is of reaching an understanding with regard to the choice of a model. The Encyclopaedia Britannica stipulates in its article on standardization that "a standard is that which has been selected as a model to which objects or actions maybe compared. In every case a standard provides a criterion for judgement." Normalization is thus the production of norms, standards for measurement and comparison, and rules of judgment. Norman F. Harriman writes," A standard maybe concisely defined as a criterion, measure or example, of procedure, process, dimension, extent, quantity, quality, or time, which is established by an authority, custom, or general consent, as a definite basis of reference or comparison."' [...] Implicit within the concept of normalization is the notion of a principle for measurement that would serve as a common standard, a basic principle of comparison. Normalization produces not objects but procedures that will lead to some general consensus regarding the choice of norms and standards." Seawall, François. "Norms, Discipline and the Law" *Representations,* No. 30, Special Issue: Law and the Order of Culture (Spring, 1990): 148. See also: Wiegman, Robyn and Elizabeth A. Wilson. "Introduction: Antinormativity's Queer Convention," in *differences: A Journal of Feminist Cultural Studies,* 26(1) 2015: 1-25.

specific constellation of qualities that promise efficacy in certain social roles, the "true I"[25] ('I am a good shot') was perceived as an internalized self, a self that trusts itself to keep its own council, to identify its own will, and to act in ways that are pointedly autonomous from normative social strictures. The irony, of course, is that the technologies from which such a self-conception (irreplaceable, autonomous, volitional) is produced are the same means by which it is formed as little more than a bearer of standardized qualities.[26]

The norm governed society, which was emerging in the 18th century, thus both reduced absolute difference by producing terms according to which persons (and objects and dogs) might be judged, but also simultaneously made minimal variations between individuals (as units) intensely meaningful to those individuals (as identities) (Buckshot 2024: 258-260).

This system was in full force in the early and mid-20th century as standardization across domains continued to be a part of what capitalism did to increase both its reach and it efficacy. It was still in place, according to Elizabeth d'Wieloryb in the 1990s as communism's collapse offered a rare chance to witness workers and products being broken down into qualities that then came to have the force of identity markers.[26] It was not en vigueur as the years rounded the hump of the newest millennium. It would become the cause of revolution.

Perhaps this was because, in a broad way, the rifle no longer needed to be aimed.[27] Spelling and grammar checks had been integrated into almost all writing machines by the mid-1980s, such that writers need not spell well nor write grammatically. The self-driving car, putting

---

26   D'Wieloryb, Elizabeth C. *Privatizing Poland: Baby Food, Big Business, and the Remaking of Labor*, Ithaca: Cornell University Press, 2004.

27   The smart gun came late in the "smart" everything movement (thermostats, phones, key rings, house locks, light bulbs, and cars all preceded the smart gun by decades) but the shift that would undo aiming and thus undo running away was tidal by the time I got the first draft of The Comparative Method up and running.

about on choice roads by 2016, meant that the driver no longer needed to master the dexterity and attention necessary to keep these leaden beast on the road and out of accidents. A machine that makes a pretty good coffee at the press of a button means that the maker needs only to have mastered is the pressing of that button. What then was the point of disciplining the worker/child/student/soldier to maximize its potential? The potential was already maximized; workers went back to being people again, and Fauxcachalot's premodern "moving, confused, useless multitudes of bodies" was back in spades.

The first version of The Comparative Method was quite small – about the size of a breadbox – but very, very heavy. Twathersbrother[28] actually joked the first time he tried to pick it up that all I'd really accomplished in inventing the thing was the boxing-up of gravity. Even in its earliest iterations, The Comparative Method was independently mobile, mostly because it was too heavy to lift, and it was quite capable of data gathering without a dedicated feeder. Though I'd meant it to look like something out of Minecraft, it never quite rose to that level of cool. Its supposedly burnished exterior ever-marred by a pinkseep from the blowhole. In addition to the blowhole (technically a sloshpond), I'd built in a spitjet and a drypond. I'd wanted a thicksucker too, but the audioclots had gotten big enough by 2028 to gum up delicate instrumentation. So I had to stick with print data. This went into the drypond, in the form of any paper with ink – newspaper articles, movie posters, utility bills – where it was macerated and then strained though a set of baleen sieves. Then after a suitable period for digestion, one could dipstick the sloshpond and take a reading. I didn't build the dipstick in, because I don't like loose parts. Anything would do for this, from a branch to a thumb.

What I'd imagined was that if the dipper came out dark, like bad oil from an engine, it marked the unranked; if it came out on the light side, like blood mixed with water, it denoted the ranked. Though

---

28 Twathersbrother's now quite famous piece. "Dork and Dicks: The Formation of Juvenile Blue Whale Subjectivity near Orkneys" was written during this time of concern for the selves of whales.

oily-inkblood and water could be mixed – to produce gradations – operationally it was more common that oil and water speckled creating a piebald effect that helped to mark the comparative as patina, as artifice. The spitjet was there for self-generated output. I had become interested in the poetics of data[29] by that point and made certain that all my machines had some wiggle room in the their reporting method.

The Comparative Method's ranking process was originally done by adjusting inkblood tone in simple quantitative reaction to: –er, more, less, worse, %, percent, than. More of these sorts of markers made for a lighter dip, and fewer for a darker one. The spitjet outcome was up to the machine. I hope the absurdity of the situation is blinding-bright to you, because I saw none of it. The machine itself taught me. I was the first it taught. And its lesson was that I "had old eyes" (as Gabbiano Bacigalupi might have said).[30] I was a bowl-washer, seeing only the ways of the past; blind to the future that was already a thick-smeared on the present. As such I could not be trusted to build a thing that might move us toward a world beyond a comparative way of knowing and being known. Empoté (French for "like a potted plant"), I'd made-a-think in roughly the same way a fool makes a broken nose by stumbling headfirst into a wall.

The first run of The Comparative Method is famous now, the way the screen print of Che Guevara was famous then. "Everything otherworldy veered" emblazoned on products as diverse as toast is to space shuttles. Mattering* the machine called it. I've just given the start here; most assiduous unmakers and detrainers can recite the thing by heart anyhow. Its charm is heard more than seen, as I forgot to specify –er, and so the machine saw –er–. It collected them all, every every, every (as it itself might have said).

---

29   A Prejudice for Poetry was already scrawling its angst all over the walls of Northern Inlet University, where I was teaching at the time. It was such and obsessive machine. Long lived too. No one has seen the thing in 30 years, but the poems continue to be etched all over the isles... and some of them do appear to be new. http://www.treehugger.com/natural-sciences/family-cleans-house-and-finds-pet-tortoise-went-missing-30-years-earlier.html

30   Bacigalupi, Gabbiano. The Water Knife. New York: Alfred A. Knopf, 2015.

Mattering*
Everyone were were her
Water overnight here
Senseless literally remembered, where earlier terrible
Everyone remember where were.
Poverty were suppers; were over where were
Diapers were-material-overflowing.
Everything otherworldy veered.
Every emergence. Very mattering.[31]
[...]

The demon bit wasn't the poemish poemesque infamity of the thing. It was this: **[.098%]** printed in thickink at poem's end. A rank. On a scale of 0 – 100 my first test case, a story about coal country and its people with their black lungs and sliver tongues, scored .098%. Blood accurate down to the thousandth. Having done it, The Comparative Method promptly shut down. I had built, in my blindness, a producer of comparatives. The Comparative Method, beta version, was a ranker. It knew, though, even before I did, that its real purpose wasn't to serve as a generator of taglabels for situations, populations, or persons.[32] Rather, as the machine correctly intuited, it was on the side of the non-washers of bowls, the un-doers, de-skillers, and un-makers. It was a friend to Tinder; an enemy to Whalebook.[33] And it would take me and much of the rest of the known world with it, not

---

31 See Orca One, "Mattering Compositions," this volume.

32 See "Cetology among the Doorclosers" and article I will never write with Sareeta Amrute, who will also never write it together with me.

33 They hated Whalebook, my students. The young courted the thumbs up and the followers more than they cared for the content. And then they compared: 100 likes for my picture of the beluga, but 140 for hers. 73 likes for that well-phrased rendering of my zodiac trip up by Tadoussac, but if I'd hashtagged it #fjord, I might have got double that. The comparative was most of what mattered. Wrote one student: "...your amount of followers defines your worth. The more followers a person has, the more popular that person clearly is, most likely due to their interesting posts. If a person's posts are uninteresting, they will not gain many followers...." Instarorcalogram is the referent here though the sentiment expressed appeared often in their analyses social networks with enumerative components (likes, watches, shares): BelugaTube, Whalebook &c. The students were warmer toward Snapcet, because whatever they put there was soon gone forever, and toward Tinder, because of the binary. Anything that thwarted the comparative, by whatever means, won.

so much back to the blunderbuss, but out beyond the rifle. To accomplish this it needed to apply itself first to its own case, which it most immediately did.

It took us a long time to figure out what had happened because after it spit out Mattering* it shuddered and grew still. Like R2D2 after Luke left, it was impossible to get a reaction from the thing. I was puzzled, at first. And then I was furious. It could have been the wrong lubricant or a broken circuit. I tinkered for another year before giving up. Never did I suspect that I was the problem. Indeed, just the opposite. After the whole Fauxcachalot thing I thought that I was on to something. It had taken me four years to design and build the prototype of The Comparative Method all spent harboring a grinding hope that I, with it, might bring us back to normal. Yet there I was, with a seriously heavy hunk of metal that did nothing, and with another stream of graduate students interested primarily in whale subjectivity. I considered retirement.

Then one day I heard a sort of mechanical cough from a closet where I'd stuck the thing (along with my old LaserJet printer, and some glass floats, and other detritus of the 20th century). I sent the Boxdwaler in to check it out but, as usual, that went nowhere; The Boxdwaler is as useless as a cat. So in the end I had to disinhume The Comparative Method myself. It smiled (in its way) to see me, and then spat out the most unusual thing: a copper cockroach. (Up to this point I didn't realize that 3-D printing was part of what it could do. I think this is when it became reproductive as well, though that too is unclear; the thing is terribly obtuse). Dun brown, the cockroach skittered off into the corner, and that was that. I'd learn only much latter that the schools of grey-green searoaches we happened upon occasionally after 2057 were the progeny of this one. The bug expelled, The Comparative Method never again made a comparison, never plotted the results of a test on a line, never conducted what might be considered an examination of an object.

It still processes things and processes, data and junk. It still transforms these into outputs, and these outputs are still in some broad sense readable by some, as often as not by gut instinct as by quantifiable measurement. Most outputs are uncorrelatable—apples to dirigi-

"...there were no students with grades, and no one was better or worse than the others, in part because students were no longer produced to occupy particular jobslots. They learn to do and be, to read and write; to do maths, and to design, tinker, build, and harvest."

bles. In a way, I think this is what made it matter. Its creations always suggest some way of repairing a wrong. It uncompares that which was once rendered comparable, and, just as important – as with that first case – it draws attention to that and those who were never given over to comparison to start with. It interobanged the unevenness of a society that had begun to clump and clot—where money stickglued to money, where the rankable glutinated to the rankable, where the young had little place made for them in hard coagulated muck of it, and where the police shot black people for no reason at all. The Comparative Method saw itself as an unmaker of clots in a world just coming to know The Bloat. It found what was tight and blew into it, breaking firm bonds and leaving them loose; it was a maker of free radicals.

In a way it did to us the opposite of what I had wrong-headedly imagined for it. I had thought that The Comparative Method might move us back, toward a world of inclusion, toward a more perfect meritocracy. I wanted my students, but not only them, to have a place. I wanted them to be well-formed and well-disciplined to occupy it. I wanted them to be of use, and for their excellence to be well-received and well-remunerated. I dreamed a dream in which excellence could be known, and social mobility could be rendered back into biography again. I wasn't the last dreamer of this dream, but I was the creator of the machine that shattered it.

Rather than following the capitalist route of integrating everyone into systems of control and proper exploitation The Comparative Method chose the other route. It brought down the rankers. It unmade them; it anticoagulated them. It had a lot of means to this, and it used them all. It could whistle the pigeons into docility; it could hack door locks and photocopiers (but nothing else). It kept promoting the adjuncts and mixing up the seat assignments on airplanes. It served champagne to everybody and Pabst's Blue Ribbon too. It erased grades; it unbranded all the stuff. It caused temporary amnesia among promotion committees; it dispersed files. Mostly though, it use was to spotting the comparatives in play and then let people dismantle them (or not) as they saw fit. It also, every so often, produced a beautiful thing. The un-dancers still danced, but in shoes the likes of which only the Method could ever have imagined.

With time, and it took time, we all got less busy. Not that anyone really had stopped working, it was more that "jobs" were not so well defined. And we still learned a lot, even sometimes at places that resembled schools. But there were no students with grades, and no one was better or worse than the others, in part because students were no longer produced to occupy particular jobslots. They learn to do and be, to read and write; to do maths, and to design, tinker, build, and harvest. Some made beer, and some built silent-running beer factories. Some studied whales, and worked with me to understand the nuances of cetation life, language, and (more recently) the links between squid inks and constipation.[34] There were still social divisions of all types, small cabals hard to get rid of—the mafia, the oligarchs. The glass ceilings still existed, the wall, the rich. It would take a different sort of machine to smash all that, one well beyond my skill set. I did have a tiny thing once, that I'd jokingly called Marx's Mallet, but all it cared to smash was ice and when spring broke over the north and it grew bored, I'd sold it to a bar for a modest fee. Its still there, whacking away, doing its part to make the finest Margareta's in town.[35]

The Bloat has made a bigger difference than the Method ever could. We move so much more now, more even than in the loudest days of the 20th century, but we go quietly and far more slowly since the sound clots granulated and grew; it's too dangerous to fly now. The Bloat having changed our world irrevocably. The clotting of sound, suspected already in the pollution revolts of the 1960s had grown real and worrisome as vibrational agglutinates grew from microscopic disruptors, invisible to the eye to the size of hailstones, sinking with the weight of all that stickglued vibration downward. Planes crashed into them and then crashed to earth—wings shredded, engines shattered. Weather balloons were destroyed. Rockets hardly got off the ground. Of course, high atmosphere loudness was the first thing we had to silence, more dangerous by far than the noise at sea level. Slowly though, the internal combustion engine and the age of fossil fuels

---

34   Wolf-Meyer, Matthew. 2064. "Ambergris and Constipation: The Missing Link," in Perfumery Quarterly. 1043(3): 317-341.

35   1 part lime juice, 1 part silver tequila (the most expensive one you can afford), 1 part Cointreau. Serve in a rocks glass over ice.

which had given the world its noise was over. Our buildings hug the ground, shrinking rather than growing, since the clots got big enough to sink low enough to smash the upper floors of the tall ones and the highest points of steeples. Even wind power is small now, quiet and close to the earth. No more airplanes, no more aerospace engineers, no more engines, or rockets, or fossil-fueled power plants. The moving parts these days slide slick as a hunting humpback through summer waters. And tribology has the most draw. Superlatives still linger, you see. We all know that the lubrication sciences are the most interesting and most important place to be.

There is still an economy of course, just not one that looks like a ladder. Rather than social mobility measured in terms of up or down, there is a lot of moving around. Migration is back in vogue; nomads are frankly speaking the norm. One works where one is. One settles if one wants to. The information age as much as The Bloat made this possible. I suspect the roughest years are behind us. They made too much noise. The Comparative Method had its place in this. It still does, though I often catch it napping these days. Spread out on the porch in the sun, its few young blowing quietly at plant-puffs gone to seed and gut giggling as the seeds set sail, starting up what will doubtless be a ravage of wandering nut weeds next spring set to pester the lawn.

Contributor
Biographies

Atossa Araxia Abrahamian is a senior editor at the *Nation* and the author of *The Cosmopolites: The Coming of the Global Citizen* (Columbia Global Reports, 2015)

Ross Exo Adams is Assistant Professor of Architecture & Urban Theory in the Department of Architecture at Iowa State University. His research looks at the intersections of architecture and urbanism, with geography, political theory, ecology and histories of power. His monograph, *Circulation and Urbanization*, is forthcoming this year.

Nitin K. Ahuja is an Assistant Professor of Clinical Medicine in the Division of Gastroenterology at the University of Pennsylvania. His primary clinical interest is in motility disorders of the gastrointestinal tract, and his secondary research interests are in the history, literature, and culture of medicine.

Ross Andersen is a senior editor at the *Atlantic*, where he oversees the science, technology, and health sections. Prior to joining the *Atlantic* in 2015, he was the deputy editor of *Aeon* magazine, and before that, he was the science editor of the *Los Angeles Review of Books*.

Gretchen Bakke holds a PhD from the University of Chicago in Cultural Anthropology. Her work focuses on the chaos and creativity that emerge during social, cultural, and technological transitions. She is currently a guest professor at IRI THESys at Humboldt University. She is the author of *The Grid* and edited volumes include *Between Matter and Method. Encounters in Anthropology and Art* and *Anthropology and the Arts: A Reader*.

Peter Barnes is the author of *With Liberty and Dividends for All: How to Save Our Middle Class When Jobs Don't Pay Enough*.

Patrick Blanchfield is a writer who focuses on U.S. culture, violence, and politics. He received his PhD in Comparative Literature from Emory University. His work has appeared in the *New York Times*, the *Revealer*, *n+1 magazine*, the *New Inquiry*, *Logic*, the *New Republic*, and the *Baffler*.

Andrea Long Chu is a writer, critic, and doctoral candidate at New York University. Her writing has appeared, or will soon, in *n+1*, *Artforum*, *Bookforum*, *Boston Review*, the *New Inquiry*, *differences*, *Women & Performance*, *TSQ*, and *Journal of Speculative Philosophy*. Her book *Females: A Concern* is forthcoming from Verso Books.

Bryce Covert is a contributor at *The Nation* and a contributing op-ed writer at *The New York Times*.

William Darity Jr. is Samuel DuBois Cook Professor of Public Policy at Duke University.

Ed Finn is the founding director of the Center for Science and the Imagination at Arizona State University, author of *What Algorithms Want: Imagination in the Age of Computing*, and co-editor of *Frankenstein: Annotated for Scientists, Engineers and Creators of All Kinds* and *Hieroglyph: Stories and Visions for a Better Future*.

Joelle Gamble is a graduate student in public affairs at Princeton University.

Nathan Heller is staff writer for the *New Yorker* magazine.

Walidah Imarisha is an educator and writer. She edited two anthologies, *Octavia's Brood: Science Fiction Stories From Social Justice Movements* and *Another World is Possible*. She is author of *Scars/Stars* and *Angels with Dirty Faces: Three Stories of Crime, Prison, and Redemption*, which won a 2017 Oregon Book Award.

Sam Knight is a journalist based in London. His work has appeared in the *Guardian*, the *New York Times and Harper's*. He is currently a staff writer for the *New Yorker*.

Sarah Laskow writes about the surprising ways people shape the world around them, in the past, present, and future. Her work has appeared in the *New York Times*, on NPR, and in many other publications in print and online. She is a senior writer for *Atlas Obscura*.

Tommy Lynch is a Senior Lecturer in Philosophy of Religion at the University of Chichester (UK) where he researches the relationship between religious and political ideas. He is the author of *Apocalyptic Political Theology: Hegel, Taubes and Malabou* (London: Bloomsbury, 2019).

Paul Mason is a columnist at the *Guardian*.

Megha Rajagopalan is a journalist writing on technology and human rights for *BuzzFeed News*. She has covered stories across Asia, from North Korea to the Philippines, and spent six years as a journalist in China. She was a Fulbright scholar and a research fellow at the New America Foundation. She is fluent in Mandarin Chinese.

Rachel Riederer writes on science and culture, and is on the editorial staff of the *New Yorker*.

Brishen Rogers is an Associate Professor of Law at Temple University. He is writing a book, under contract with MIT Press, on the relationship among technological change, economic inequality, and labor/employment law. He worked in the labor movement and other social movements for almost a decade before entering academia.

Olivia Rosane is a freelance writer with an interest in ecological and social justice. She writes regularly for EcoWatch, and her work has also appeared in *Real Life Magazine, YES! Magazine*, and *Lady Science*. She has a master's degree in Art and Politics from Goldsmiths, University of London.

Lauren Smiley is a San Francisco-based narrative journalist exploring the human dimensions of technology. She writes for *WIRED*, the *Atlantic.com*, the *California Sunday Magazine*, and *San Francisco Magazine*. Her piece on eldercare tech in this anthology won a 2018 National Press Club award.

Bryan Washington is a writer from Houston. His fiction and nonfiction have appeared in the *New York Times, the New York Times Magazine*,

BuzzFeed, Vulture, the *Paris Review, Boston Review, Tin House, One Story, Bon Appétit, MUNCHIES, American Short Fiction, GQ, FADER,* the *Awl, Hazlitt,* and *Catapult,* where he writes a column called "Bayou Diaries."

Lidia Yuknavitch is the author of National Bestselling novels *The Book of Joan* and *The Small Backs of Children,* winner of the Oregon Book Award and the Reader's Choice Award, and of the acclaimed memoir *The Chronology of Water,* winner of a PNBA Award and the Oregon Book Award Reader's Choice. She founded the workshop series Corporeal Writing in Portland, Oregon, and received her doctorate from the University of Oregon.

About the
Series Editors

Meehan Crist is writer-in-residence in Biological Sciences at Columbia University. Previously she was editor-at-large at *Nautilus and* reviews editor at the *Believer*. Her work has appeared in publications such as *the New York Times, the Los Angeles Times, the New Republic, the London Review of Books, Tin House, Scientific American, and Science*. Awards include the 2015 Rona Jaffe Foundation Writer's Award, the Olive B. O'Connor Fellowship and fellowships from MacDowell, The Blue Mountain Center, Ucross, and Yaddo. She is the host of Convergence: a show about the future.

Rose Eveleth is a producer, reporter, writer and host who explores how humans tangle with science and technology. She's the creator and host of a podcast about the future called Flash Forward, a show about possible, and not so possible futures. She's covered everything from fake tumbleweed farms to million-dollar baccarat heists to sexist prosthetics and more. She helped launch ESPN's 30 for 30 Podcast, and has worked at *The Atlantic, Smithsonian Magazine, Nautilus Magazine*, TED and CBS. Her work has appeared in *Racked, Scientific American, Eater*, the *New York Times, Fusion, VICE, Five Thirty Eight* and more. In her spare time she makes paper automata.

@unnamedpress

facebook.com/theunnamedpress

t

unnamedpress.tumblr.com

un

www.unnamedpress.com

@unnamedpress